A Mutant Ape?

The Origin of Man's Descent.

Published by Merops Press

websites:
www.cosmicconnections.co.uk
also www.scienceandphilosophy.co.uk
www.scienceandthesoul.co.uk

Acknowledgements: Suzanne, my wife and Françoise, my daughter

Contents

Illustrations

1.   For Starters

Ape > man or ape | man? What, however, is the real issue? It is, first and finally, holistic versus materialistic interpretation of the evidence. It is, although much protest and confusion is raised, debate between atheistic and non-atheistic positions.

Your origin defines you. Therefore, are humans the beneficiaries of cosmic order planned for us to flourish or the offspring of blind chance? No doubt, each side in this fundamental debate wants honesty and truth. Be forewarned, however, that the ape-man arena resembles a thrust-by-thrust duel generating the sparks and dusts of explosive confrontation; and that the space is policed by 'authorities' of either party who make it their religious business to prevent the other conviction 'winning'. Such authorities present their arguments in many ways. Essentially, however, in contemporary scientific culture reason is declared materialism's friend; and belief in *natural* reason (natural teleology or immaterial cause) is deemed irrational and thus, by any means, an antagonist to be dispatched.

'*AMA?*' can be read in its own right but, more profoundly, plugs umbilically into *SAS*. In other words, this book expands on a short section called 'A Mutant Ape?' from Chapter 25 of 'Science and the Soul'. Consequently, it assumes the same 'wrap-around' framework as *SAS* viz. a comparison of materialistic and holistic interpretations regarding the historical origins of life on earth - including yours.

For new readers a brief introduction to this linkage and its framework follows. This is necessary in order to engage fully with the thrust of the book. Those already familiar with Natural Dialectic can recapitulate or, on the other hand, flip to Chapter 2.

The anti-parallel perspectives[1] that govern such narrative are called *top-down* and *bottom-up*.

Top-down? In physical nature *per se* there is no such thing as a number or a name. There is no symbolic code. Matter is oblivious. Mathematics, language, logic, reason and so forth, derived by conscious mind, are 'unnatural', psychological factors. They constitute an immaterial, metaphysical element called information. Information

---

[1]   see Glossary; also *SAS* Chapter 0: Two Pillars of Faith: three further books fully expand the theme - A Potted Grammar of Natural Dialectic (*PGND*), Science and Philosophy - The Lectures (*SPFP*) and The Reunification of Science and Philosophy (*RSP*); also endpage and website: www.cosmicconnections.co.uk.

always *means* something. Its source is mind, it dwells in mind and musters purpose, knowledge and an orderly construction.

Holism's *top-down* gambit adds to this immaterial element the mindlessness of all material things. Information occurs in both active and passive modes. The passive involves both subconscious and material expressions of active, informant consciousness. Thus, the *top-down* perspective ranks information above matter. Information is prior; energetic patterns depend on it. For this reason *SAS* has expanded on the materialistically heretical notion of 'archetype'.[2] Such informative notion *includes* physical mechanisms expressing variation, speciation, mutation,[3] so-called micro-evolution, natural selection[4] and limited plasticity. **Why should such plasticity not apply to human as to other types of organism?**

Conversely, *bottom-up* materialism's gambit is that every object and event, including an origin of the universe and the nature of mind, are material alone; a few oblivious kinds of particles and force compose all things. Moreover, cosmos issued out of nothing; therefore, beyond this realm of physics there is only void; and life is an inconsequent coincidence, electric flickers of illusion in a lifeless, dark eternity. It ranks energy above information; psychological patterns are physically created.

This view's necessity, evolution, is materialism's *sine qua non*. As such its story (which explains your chancy origin as though a fact) is defended to extreme degree by religious materialism, that is, a mind-set called scientific atheism.

In this way philosophical presumption conflates two gambits into one. It assumes that informative consciousness is an 'outcrop' of evolved brain chemistry: it guesses that patterns of mind are, essentially, an illusion of oblivion! By such paradigm, of course, no dualism of separate mind and matter exists. Information and energy, mind and matter are of wholly the same substance. And if this substance is material it is non-conscious. Consciousness is therefore a peculiar effect of certain formulations of non-consciousness. How strange are mind and subjectivity! How queer 'scientific' speculations made by quarks and leptons of a brain!

Your body is certainly made of cells, cells are made of chemicals, chemicals of atoms and atoms are not, any more than quarks and leptons, subjectively alive. If atoms, molecules and cells aren't living

---

[2]   see Glossary: archetype; also *SAS passim*.

[3]   see *SAS* Chapter 23: The Creator or G.O.D (generator of diversity).

[4]   see *SAS* Chapter 22: The Editor.

then your whole body isn't. It might be a marvellous machine but it is thoroughly non-conscious. *So who are you? Are you alive or dead?* **It follows that a scientific world-view that does not profoundly and completely come to terms with the nature of conscious mind can have no serious pretension of wholeness.**

It is important to mention this issue, dealt with at length in *SAS*, because it sets the tone for 'A Mutant Ape?' No doubt the metaphysical notion of an archetype[5] defies materialistic common sense. Equally, however, the notion of random, mindless evolution of highly informed (bio-)systems defies intellectual sanity. *Is it logical or reasonable to proclaim that complex code, the symbolic purveyor of meaning and purpose, could aimlessly evolve?* Random engineering of bio-evolutionary code and its dependent product, bodies, is as seriously irrational a theory as a grasp of 'informative bio-archetype' is rational. **It might, therefore, be added that a scientific world-view that does not profoundly and completely come to terms with the nature of archetype can have no serious pretension of wholeness either.**

And so to apes and men. Did man evolve or is he archetypally designed? Theists of distinct colours each allow one of these arguments. Either a 'remote' deity set an externalised creation rolling on a random, evolutionary course[6] or there is straightforward design by an agency that projected itself, as we do into our own work, into the creation of bio-forms. But an atheist has no alternative. There is, for this group, no start-up intelligence, no natural teleology, no specified design. Information[7] is defined in objective terms of statistical probabilities. Therefore, the immaterial, psychological dimension of universal mind and archetypes is thoroughly excised from all possibility and, thus, from all discussion. Indeed, the group's gladiators fight tooth and nail to preserve its creedal heart, the Darwinian theory of evolution according to which codified life-forms are the chance product of natural forces, that is, of mindless matter alone. Is this theory a fact? Or is it due for radical revision? This book will help to show.

If, two hundred years ago, I had revealed that you were made of billions of tiny cells, each exquisitely functional and containing, in a nuclear book of instructions much too small to see but thousands of times longer than any holy writ, a symbolic reflection of your total form, you might have scorned me as a madman. The meaning of all books is reflected in symbolic code called language; and each of your nuclei symbolically stands, as regards its *DNA*, for you. **You, at least**

---

[5]   intrinsic habit; typical, subconscious centre of influence: *SAS* Chap. 16.

[6]   theistic evolution - see *SAS* Chapter 24: Theories of Accommodation.

[7]   see Glossary and *SAS* especially Chapters 5, 6, 16 and 19.

in body, are its meaning. *And, if computational code expresses intent, I cite psychological computing in the form of archetypal storage, referencing and reflection of intent as the source of humanity* - to a sneer of disdain and modern, materialistic thumbs-down!

Yet the fact is, nothing makes sense in biology except in the light of immaterial information. You may drag evolution in on information's tail but it is simply a fashionable word[8] used in biology to mean several different things. It is not, trivially, the origin of species but, fundamentally, the origin of information that is at issue. **Not evolution but symbolic information is the basis of biology. Thus the subject's basis is a non-physical entity. Its root is metaphysical.**

It is, moreover, against all known laws of physics that oblivious matter might <u>create</u> *information; or that material instruments of purpose self-organise or instigate their own programs.* **There is neither law nor known agency that can cause information to originate by itself from oblivion, that is, from matter.**

We can proceed further. **A computer is a mind machine. Programs are a mind machine's intent. They express the will that mind invested in machine.** *It needs be re-emphasized, every machine (including body biological) has the mind of its maker in it.* Not in its atoms *per se* but in its purpose, design and lawful operation. Machines passively embody information. Mind-machine information is passive. Active has produced passive information. **There is, to emphasize the fact, known neither law nor process of nature by which matter originates information. The source of information is always mind. And at the heart of mind is consciousness.** Such common sense sounds ridiculous to the counter-intuition of an evolutionist who has denied it in the first place!

Materialism's problems don't end here. While a computer uses binary digits (0 and 1), functional bio-information is coded by four *DNA* digits (cytosine, guanine, adenine and thymine). A key prediction of Darwinism is that, especially as regards transformative macro-evolution, functional information content must increase from zero until a first cell[9] is 'self-organised'; after this net increase over time continues until at least the contemporary diversity of life[10] occurs. Such macro-evolution's source of new information is presumed to be random, genetic mutation. **But in fact mutations, creators of mutants, embody entropy of information; the relentless, net effect of random mutation is degradation unto destruction of function.** Entropy does not create; its direction is, for

---

[8]   check Glossary.
[9]   see *SAS* Chapters 20 and 21.
[10]   see *SAS* Chapters 22 to 25.

life's Titanic, sinking (↓) down. Thence it might be argued that mutation, as an explanation for macro-evolution, is an unproven and, if reason is set dead against coincidence, mythic system of belief.

The holistic prediction *is* contrary. Serial, random perturbation of integrated networks (with which life-forms are replete) does not lead to novelty but destruction. Information degrades with time. One thing every programmer strives to eliminate (with an effect proportional to his intelligence) is bugs, error and the chaotic part of chance. Therefore, you would predict full functional information content at the start. After this, despite editorial systems embedded to slow the process down, net degenerative bias would obtain; entropy of information would gradually garble than delete a program's operative sense.

What is found? Such retrogressive bias has already been found in bacteria, fruit flies and humans. And while net deletion wins the day macro-evolution has never been seen to happen.[11] The Darwinian theory of code creation runs against the cosmic grain; theory-driven macro-evolution, where little bits of digit-change can accumulate into creative, fully codified and complex innovation, flouts the real world's entropy. Indeed, chockful of gaps it simply speculates that from unidentified ancestors by undemonstrable and disputable means over uncertain lengths of time one species of 'missing link' begat the next. Guesswork pervades at every step hypothesized beyond the mundane triviality of variation-within-type. Why, therefore, do so many institutions believe it? *It is both because the naturalistic methodology of science[12] demands materialistic answers; and because the mind-set of atheism, a creed that unfurls 'scientific' across its banner, demands that such a creation story **must** be true.* Reputations, specialist careers and, above all, a world-view depend on it.

So, in summary, *bottom-up* evolutionary theory asserts a random origin for humankind. This assertion is assumed to be a fact. The norm is thus for each relevant discovery to 'rewrite history' in a continual revision of the truth. Such adventure is the order of materialism's scientific day.

On the other hand, *top-down* holistic theory will assert that the same theory of evolution is today's phlogiston. And its oxygen is information. Modern understanding of the world we live in must, as its foundation, sensibly accept the element of immaterial information; and universal information's 'atom' is an archetype.

---

[11] *SAS* Chapters 22: Galapagos and All That and 23: Evolution in Action?
[12] check Glossary.

2. *Troublesome Irrationality*

Illogicality runs deep. The world, for a materialist, is basically colliding particles and interactions due to force. The energy of this world does not change but is transformed in patterns physics formulates. Biologically, however, patterns of transformation are not informed by blind force but by symbolic code. Nowhere, either outside or even inside cells, does that slender thread on which materialism hangs, a 'self-replicating' codifier, exist. Indeed, no chemical at all can magically double up. In fact, hundreds of different molecules, each precisely codified, cooperate to copy and thus replicate strands of so-called macro-molecular *DNA*.[13] Nor even then does such nucleic acid know or care a fig; what target does a bunch of atoms ever have? Such oblivion is also ignorant of 'ought', morality, laws of logic, reason or intelligence. Even *if* its carelessness evolved development of nerves (another great guess in doctrinal dark) why should their chemical reactions produce a calculation, a sonnet or a thing of beauty? Indeed, why should one set of atoms (your brain *aka* you) understand another (the rest of your body and its universe)? And how can a set of physical changes, physically caused, possibly 'correspond' to such conscious experience as seeing that 'an axiom is self-evident' or to conscious logical transition as implied by the word 'therefore'?[14]

Beyond the chemicals that make it up is any machine purposive? Of course, if you include its engineer. But, unrealistically, materialism eschews purposive design in nature because 'there is no engineer'. And so the fluttering of a wing 'appears' designed but is 'not really'.[15] Equally, thought in evolved nerves only 'appears' logical or innovative. The flutter of logic is, apparently, as illusionary as the steps of purpose. Thus any academic case, including Darwin's, may 'appear logical' but actually is not.

To rephrase: if a whole sequence of logical steps, mathematical algorithms or reasonable statements were indeed merely the effect of a causal chain of physical processes, all blindly and mechanically or electrochemically determined, it would follow that the speaker could not think what he wanted or help saying what he did. He would, quite literally, not know what he was talking about. His statements, as

---

[13] *SAS*: Chapter 20: Perplexity.

[14] *SAS*: Chapter 3: Hierarchical, Triplex Construction of the Cosmic Pyramid.

[15] *SAS* Chapters 3, 4 and 6: Machines.

reasoned arguments, should therefore carry no weight. *Thus, why should we believe a word he says*? **In this respect it is common knowledge that, on his own terms, a serious atheist is talking gobbledegook**. Physicist Sir John Polkinghorne rams the point right home. 'Thought is replaced by electro-chemical neural events. Two such events cannot confront each other in rational discourse…The very assertions of the reductionist himself are nothing but blips in the neural network of his brain. The world of rational discourse dissolves into the absurd chatter of firing synapses.'

In this case what's the theory of evolution worth? If it proposes that human mind is a product of randomness constrained by natural selection then survival value and not truth becomes the criterion by which it judges what is right and true. And, since survival is *per se* no guarantor of truthfulness, why should mind's theories value truthfulness? By this yardstick we cannot tell a true idea from false. Why, therefore, shouldn't evolutionary ideas be wrong not right and therefore worthless? Can the theory logically survive itself?

If, moreover, as some naturalists aver, consciousness is an illusion of grey matter what are concepts such as 'free will', 'moral order', 'sense of self', 'God' or 'no-God'? Are they as false as, consequentially, reason, rules of logic and mathematics? Theorising can't be trusted either, any more than grand conclusions that concern the 'origin of man' or 'theory of evolution' in entirety. After all, you couldn't trust computers built irrationally. Why, therefore, trust the unpredictable illusions of a mind-brain built the same - unless you think oblivious force and chance are rational. Irrationality does not make rationality but saps it. Oblivion cannot innovate; chance can't follow an idea. And clever bodies, irrationally constructed for some 'reason' by survival to survive, can't by this criterion distinguish an unchanging truth or error from adaptive ones. Their type of reason, whose objective isn't truth, undermines itself. Atheism similarly, by adherence to the creed of evolution, similarly undermines itself. Why should we believe a word?

So let's get rational. Let's look askance at nonsense and beware absurdity - not least when its enchanting paradigm, materialism, might well spell trouble as it leads astray.

3. Skew

When your car tows to one side the garage diagnoses a tracking problem. Although you may have got used to it, the problem is real and, by balancing repair, set straight. But the fact is that, by its methodology, by the framework within which its answers *must* be rendered and by an *a priori* exclusion of any metaphysical factor, science as a venture is invariably and incurably skewed. *Such closed-minded skew rests on assumption and may be philosophically illogical but, like it or lump it, in general it constitutes the bottom-up, scientific bent of mind.*

It is, therefore, inevitable that materialism's authoritative story of our origins, the Darwinistic creed, also incorporates invariably and incurably present tow or skew. Tautological skew. In rationalistic circles evolutionary theory is taken for granted. It is construed as unquestionable fact because, materialistically, it *must* be true. And our world is so instilled with this repetition that only such a narrative has become acceptable, only such lens allowed. Thus holistic balance by the garage is proscribed. Inherent imbalance, according with materialism's bent, permeates interpretation of all prehistoric facts.

Science requires, in principle, repeatable experiment to verify a theory. With evolution, as often with cosmology, what happened in the past is neither observable nor repeatable. Not precise experiment but best guess (called abductive reasoning[16]) therefore has to win the day. And if the questions and the answers that frame interpretations of bare bones are materialistic alone then what you'll get, output from input, is the kind of story that you're looking for. Darwinian evolution has been injected into the world's thought-stream to the point of rewriting the story of life on earth. Thus human history has also been revised through, as we'll see, the narratives of Darwin (without much precise reference to fossils), Huxley (without ever seeing the Neanderthal skull), Boule, Dubois, Piltdown, von Koenigswald and, later, fossils out of Africa. **We'll look at these and more but, upfront, note that all discoveries have, without exception or regard to fit, been shoe-horned into evolutionary skew.**

'Ass is taken for a man' (Daily Telegraph, 14 May 1984, p. 16). A skull, found at Orce in Spain and promoted as the oldest example of *Homo* in Eurasia, was later identified as that of a young donkey. In this case it seems man *was* an ass! And in an article called 'Humanoid

---

[16] see Glossary: logic.

Collarbone Exposed as Dolphin's Rib' Tim White, an associate of Don Johanson, accuses a fellow anthropologist of *faux pas* on the scale of *Hesperopithecus* and Piltdown man (see later).[17] He does so because the bone in question is not properly curved and has a tiny opening, called the nutrient foramen, opening the wrong way for an ape-man. White has written: "The problem with a lot of anthropologists is that they want so much to find a hominid that any scrap of bone becomes a hominid bone."

A further problem is that scientific evidence for *and* against evolution is not countenanced because of a closed, sub-academic refusal to recognize that the latter exists; or to allow that *both* interpretations of the biological facts are possible.

Thus materialism's skew is equally and unconsciously embraced by research fellows, journalists and laymen. For example, the periodicals Nature, Science, New Scientist, Scientific American and most science classrooms world-wide combine to generate a powerful, pervasive and continual slant. And in the court of orthodoxy a critical objector is, as happens with heresy, at least frowned upon, ostracised and sometimes execrated. The assumption is, because 'evolution is a fact', that man *must* have evolved from apes. That's what we're locked into looking for, that's what theory says we'll find. Whole libraries, museums and institutions are devoted to this narrative. Ape-men. Our conceptual world is (though none of them apparently exist today) built around their being. Is, however, what you conceptually take for granted thus necessarily true? Or, in this case, barren? Will the skew turn out to be the straight path or a way astray?

Let's explore the strengths and weaknesses of such determination.

---

[17] Anderson I. 'Humanoid Collarbone Exposed as Dolphin's Rib', *NewScientist*, vol. 98, no.1355, 28 April 1983, p. 199.

4. *Imperative*

Man *must* have evolved from apes. 'Must' is an imperative that breeds its strategy and tactics. Strategy's to prove it true. Tactics must protect, prop and promote ape-man hypotheses until they cannot be denied.

Proof of evolution in the truly scientific sense is not available. Exact events in history happen only once and if, as basic to the theory, they are due to unexpected chance or circumstance, then they are experimentally unrepeatable. Nor is a mass of evidence exclusively interpreted one way alone necessarily correct in that construction. A hidden factor may, when realised, evoke correction. Nor, finally, is a theory that cannot be experimentally disproved a scientific one. Thus, as all detectives on a case well understand, you best-guess by interpreting the facts within the framework of your knowledge and experience.

Informative creation (whose foundation is the source of information) and energetic evolution (whose foundation is a cosmos made of natural force alone) are the two basic theories of prehistory. *In fact, neither a theory of historical origins based on the inference of chance nor one based on informative competence is testable and thereby falsifiable by naturalistic methodology.* **In this respect neither is scientific. <u>Both should be labelled pseudoscientific</u>**.

Therefore, non-experimental objectivity's the goal. How, though, is this achieved when the metaphysical skew of a mind-set can continually knock it off course? For example, today's fashionable imperative is framing a materialistic explanation of events - in this book's case human evolution. Since atheism's only cogent form of opposition is theism was it Adam or a monkey's evolution bred you? The way that we arrange the fossils proves, doesn't it, that there *were* ape-men? Death, therefore, to created Adam since proof naturally selects him out - unless the evolutionary selections made of swapping branches, guessing missing-links and arranging skulls prove incorrect! Could they? Will we find another answer in this book?

Either way, when it comes to primal parentage, sustenance of chosen theory is a philosophical imperative. Rock-hard mind-sets fight the good fight and, as any self-respecting politician knows, there are many ways to deal with opposition parties or with facts ill-fitting to his ideology. We'll come across some tactics that have been employed extensively, sometimes obsessively. In the governing, evolution party these include denial, disregard and suppression (by omission or negative commission) of any widely-aired debate about the facts.

There's also reclassification, changing estimates of dates to better 'fit' theory and manipulation unto complication, confusion and sophisticated levels of uncertainty regarding what is construed as relevant material. On the positive side, promissory notes are issued that kick required evidence for assumptions into the future or propose as yet undiscovered materialistic, mindless processes which mimic teleology. You'll find solutions in the promised land! On the negative side, you find weapons such as aforementioned tautologous argument by 'begging-the-question', straw-man demolition, *ad hominem* polemic and other debating ploys; even fraud and straight deceit have been employed.

Things change. Academics disagree. Spats in the party may attract avuncular displeasure but the really unpopular alien-outside-the-family is a non-evolutionist. Expulsion or exorcism at the fiery, intellectual stake are meted out to heretics in order to satisfy the imperative of evolutionary scholarship and naturalism's royal authority. Interpretation of the world's text *has to be*, most philosophers of science can agree, a wholly naturalistic one.

Scientific method thrives on the adrenalin of doubt - except, it seems, where atheism's central tenet is concerned. In this the party is united. 'We are absolutely right.' And men who think they're absolutely right promote their truth. Therefore, which common, powerful promotional tactics have treated or still treat evolutionary news?

One is unremitting propaganda by mass media. If indoctrination brooks no rival then, after a while, its audience (in this case the western world) succumbs. Peer pressure forces loss of critical faculty. The mind-dead mode of thought becomes habitual. This, we shall see, has happened for over a century and continues apace.

How did it start? We'll deal later with the early bones recruited in a battle to win hearts and minds. What next propels the propaganda train?

5.   Cartoons

5.1   *Hesperopithecoids* (man and wife).

A picture's worth a thousand words. Or, as Stephen J. Gould wrote in his book Wonderful Life, 'The iconography of persuasion strikes even closer than words to the core of our being. Every…advertising executive has known and exploited the power of a well-chosen picture'.

Artists have always been on hand to flesh out sparse bones with evolutionary assumptions and imaginations. Only a thigh bone, skull fragment or a few teeth have inspired reconstructions of brutish, naked, wild-haired ape-men with their signature, a stupid-looking gaze - not civilised, you understand. Indeed, on receiving a single tooth, found in Pliocene deposits by a Mr. Harold Cook, H.F. Osborn (the first President of the American Natural History Museum to have been trained as a scientist) declared it had characteristics which were a mixture of human, chimpanzee and *Pithecanthropus.* He named it *Hesperopithecus haroldcookii* (Harold Cook's Evening Ape). In Britain Sir Grafton Elliott Smith FRS, FRCP fully supported him. Thus, on the basis of a single tooth, there sprang fully-formed across a centre-spread of the *Illustrated London News* (24 June 1922) an artist's impression of the brutish *Hesperopithecus* cavorting beside his ugly, squatting wife - half animal, you understand. Later investigation proved that the tooth was that of an extinct pig. So a pig made a monkey out of evolutionists, though scarce publicity was given to that fact; although discredited, the original misinformation continued unchecked to insinuate the public mind.

Thousands of such fantastic portraits have been painted and then naively, reiteratively and unquestioningly printed in millions of copies of journals, magazines and books. Serious palaeoanthropology has not discouraged such frivolous pursuit.

Zinjanthropus (*Paranthropus boisei*) and KNM-WT 17000

  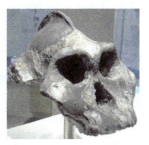

Zinj x 2 and *Paranthropus aethiopicus* (The Black Skull)

 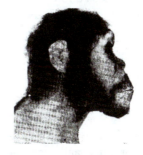

Zinj enhancements by Maurice Wilson (BMNH)

A clutch more (of many) imaginative Zinjes

5.2   Variations on Zinj (*aka* 'Dear Boy') Theme.

For example, in the early 1960's a skull dubbed 'Zinjanthropus' and, more humanly, 'Nutcracker Man' was reconstructed from about a hundred pieces. This African character, perhaps a gorilla judging from its sagittal crest, was honoured with many fantastic and very different mug-shots - for example, for The Sunday Times, The British Natural History Museum, The Smithsonian, Dr. Kenneth Oakley and similarly (looking almost human) for The Illustrated London News and The National Geographic. We'll meet Nutcracker Man and Kenneth Oakley again later.

Theory resurrects a few bones wondrously. Accordingly and more recently Dutch twins Alfie and Adrie Kennis have jumped aboard the media train. These and other modern 'palaeoartists' interpret the 'chain of early hominin species' using various transitional series of drawings, paintings and truly captivating three-dimensional anthropological models that, although more man-like than they used to be, are still susceptible to a hypothetical version of ancestral reality. And even this is dependent on questionable reconstructions of bone fragments, serial arrangements of skulls and the interpretative will to realise an ape-man. Is it possible the creature is a child of theoretical desire?

Progress in nature? Mindless nature is conservative. Nevertheless, minds see progress and 'graduated progress'/ 'serial improvement' from A to B, C and so on is a motif intrinsic to evolutionary theory. Thus it's noted that, of course, chimps and humans show similarities. You can variously arrange size or shape sequences of skulls or bones. If you *assume* that similarity indicates common physical descent (while for similar models engineers assume common *conceptual* descent) then a graduated, textbook line of 'morphs' from ape to man must confirm your theory.

5.3   Clark Howell's Deceptive Parade.

In this vein perhaps, of all propaganda tools, the most powerful and far-reaching effect has been wrought by an advertisement for evolution called Clark Howell's 'March of Progress'. Francis Clark Howell is considered by some the father of modern palaeoanthropology. His parade, drawn by Rudolph Zallinger, appeared in the Time-Life Nature Library series Early Man (1965 ps. 41-45). Recognised by hundreds of millions world-wide, it shows a single file of walking hominids becoming increasingly upright, tall and hairless. The fifteen range from skipping chimp to six fully upright figures that include examples of so-called *Pithecanthropus,* Neanderthal, Cro-Magnon and modern men and all labelled *H. sapiens. But the smooth, beguiling, progressive transformations are a straight illusion.* Even at the time of printing scientists knew that it was visually misleading. On p.41 Howell wrote that, 'Although proto-apes and apes were quadrupedal, all are shown standing for the purposes of comparison'. This mischievous slant means that at least the first five non-hominids and most probably the first seven characters should not be standing!

5.4   Origin Stories

Now, man at top, where's humankind evolving next? It seems as if the time-line's peaked. Has 'bio-progress' stopped and, if so, why? Of course, conceptual and technological advance occurs. And instances of moral progress show the best of human nature but has the general behaviour of our species much improved? Tied to physical desires, how far from its worst potential - pride, greed, anger, cruelty or lust - has mankind's mind-set as a whole evolved? *These are psychological affairs but if there never was an ape-man, only archetypal man, then bio-evolution's march through physicality is an illusion anyway.* Howell's scene pops like a bubble leaving you still here.

5.5   Human skull sequence arranged from smallest to largest.

Although cranial capacity (or brain volume) has historically been associated, the larger the greater, with intelligence and complexity of society this idea of progression is simply incorrect. Bees and ants have complex social lives but what about their brain size? Is female of the species less intelligent than male, pygmy than westerner, mouse than elephant or octopus than whale? Humans of equal intelligence are born with a variety of skull sizes; and different levels of intelligence shine from same-sized skulls. The relative stupidity of fossil man, rising to enlightened you and me, is perhaps a well-rehearsed, evolutionary-tailored myth.

Similarly, the fictional nature of Howell's iconic fresco was publicly admitted only 35 years later.[18] **But such was the impact of such fake parade's simple, brilliant propaganda that the deed (or damage) was done. It lingers falsely embedded in the modern mind-set.** Indeed, it is retained as an icon of upward mobility. And the whole, ever-shifting confabulation of interpretation and the regular discovery of new fossils keep the pot at boiling point; they combine to complicate s by this method. For example, some Australian H. erectus bones have cry out for equal research from an infidel, non-evolutionary stance.

At any rate, pot-boilers and not sermons are what keep the public's interest up. Faith, like any fire, needs feeding. As with pictures so with literature. In journals, magazines and books it is common practice to present evolution in an acquiescent, glowing light. Scientific authority dictates what is to be believed. Lack of evidence or answer isn't mentioned until a plausible example's found. Straightway such new discovery is trumpeted as if it solves (or nearly solves) the issue and if, after a few weeks or months, dissent or disproof occurs the message is quietly dropped. Depending on how cryptic or low-key a retraction the general public is left with an impression that the case still stands. Anyway, another evolutionary revelation soon appears. Continual boosters, as often as every few weeks regarding one aspect of Darwinism or another, immunise against the bugs that holistic interpretations represent and keep non-specialist lay persons thinking that some final ex-Darwinian answer's near.

---

[18]   Nature 403: Vol. 27 p.363, 2000.

6.   The Imperative of Law

Pictures and the press might spin the truth in legal ways. Even more persuasive for your cause might be the law itself.

In 1922 William Jennings Bryan, a successful politician who had in 1896, at the age of thirty-six, been nominated Democratic candidate for the Presidency, was campaigning in the courts against children being taught in schools that they were descended from apes.

The idea of the Scopes Monkey Trial of 1925 in Dayton, Tennessee, seems to have been hatched in New York by officers of the American Civil Liberties Union (*ACLU*). The legal defence, which hired famous criminal lawyer Clarence Seward Darrow, was arranged and paid for by the *ACLU* and members of the American Association for the Advancement of Science (*AAAS*). The *ACLU* released to the Tennessee newspapers a call for a teacher who would break the 1925 state law against teaching evolution. The plans for John Scopes, a football coach and substitute science teacher, to be the defendant, were made at an informal meeting in a Dayton drugstore.

Whether or not he actually violated the anti-evolution law, Scopes was indicted and W.J. Bryan came down to act as special prosecutor. Darrow, from Chicago, acted as chief counsel for ·the defence. Two main lines of evidence for evolution were the Piltdown man and (see *Cartoons*) Nebraska man *(Hesperopithecus haroldcookii)*. This imaginary creature, spun from dental floss, was heralded as the first higher primate from North America. Nowhere in the trial did the scientific problems receive any more sensible discussion! Darrow displayed ignorance both about the theory of evolution and the teachings of the Bible, and levelled a barrage of insult and vilification at fundamentalist Bryan. Bryan did not respond in kind. Darrow was clearly the media favourite, however. Scopes was convicted of violating the Butler Act and fined $100. His conviction was overturned by the State Supreme Court. Bryan, while resting in Dayton after the case, died. Creationism had been crucified. And, as the ACLU wished, the public was educated on evolution.

Such monkey business set the foundation, if not the tone, for public education and drilled, locked-in acceptance of the theory of evolution as a fact. You are, the narrative insists, a mutant ape.

It is the same story today. The *ACLU* and fundamentalists are still locked in mortal combat: and bones still rattle and roll. Evolution and creation - both should be labelled pseudoscientific. *But labels don't decide reality.* Scientific or unscientific, did serendipity or logic generate our world? Instruments can't measure supernatural creativity - but we *can* infer creation by intelligent designers, ones called engineers. **The**

issue is not one of identifying a particular designer but detecting such criteria as spell design or lack of it. You might, for example, expect to find in a design purpose and such rapid infusion of information as elaborates a concept and transforms it into a functional system; also complex integrated parts, possible use of symbolic code (such as facilitate on-off or more complex switching systems) and the re-use of functional units (or mosaic variations on their theme) in different and physically unconnected systems. *You find all these substantiating what we call biology.* **In fact, the priority of innovation is, as it involves the immaterial component of technology, information.**

From the detection of such factors one might predict biological discoveries and, thereby, infer a generator of innovative information. This, we've seen and will see, cannot be rationally identified as chance. So why black-mark, cane and then expel the rational alternative? In what kind of school does 'scientific reason' thus irrationally master us?

Holistic *predictions* could be confirmed by the following discoveries:

1. intricacy, coordination, complexity of coherence, specificity and an irreducible number and arrangement of parts per functional bio-mechanism.
2. reasons *why* specific, predetermined parts are co-assembled, that is, *why* as opposed to how they self-organise in cybernetic order to perform a task.
3. in accordance with conceptual biology (*SAS*: Chapter 19) a convergence or mosaic appearance of modular parts.
4. the discovery that previously misinterpreted components are not in fact vestigial junk but structurally and/or functionally necessary.
5. that a functionally-integrated system involves the complex operation of language, that is, accurate symbolic code.
6. that, if discrete systems are designed (albeit with intrinsic, predetermined adaptive potential to flexibly respond to circumstance and thus retain the balance to survive), you would also expect to find discrete types of organism.
7. accompanying such discreteness a case of fossil 'abruption', that is, an historical lack of millions of graduated missing links.

The presence of all such factors in life on earth might lead a rational man to infer its design. And, based on this perspective, propose such research fields as the computer-like properties of *DNA*, non-protein-coding (n-p-c, see Glossary) genomic function, front-loaded adaptive potential as regards both environmental adaptation and development, the practicalities of applied protein engineering and so on. A reductionist is, on the other hand, forced to identify his 'designer' as chance a-dance with death; and to narrate hypothetical stories and suggest research objectives accordingly. Does current law uphold discussion of such rational imperative in schools and universities?

Instead strange legal logic runs amok. Groups lobby parliaments and make petition to Supreme High Courts. Should they find in favour of 'design' or Design? If such a Court wishes to rule on a case of 'unconstitutional violation of principle of the separation of church and state' which church should it choose? Should it admit each species of faith, neither or simply an opinionated 'chosen path'? Which path? Which guess? Today the purveyance of evolutionary theory is institutionalised; but doesn't a belief in such naturalistic authority create atheists as logically as a belief in design makes theists? If theistic realism were denied and the atheistic position permitted to partake a monopoly of 'legal protection' then, as Harvard graduate, law clerk in the US Supreme Court and long-time legal academic at UCLA, Phillip Johnson, writes, 'the Supreme Court will in effect have established a national religion in the name of First Amendment freedoms'.

It's not only Uncle Sam! Nor only judges but our European politicians also seem to know it all! This Goliath totally ignores (or never learned) that the Renaissance, progenitor of modern scientific method, derived enlightened sense of order from a universe intelligibly created by a Most Intelligent Creator. In 2007, however, at once dismissive of its founder scholars and the strength of argument from inference of 'intelligent design', the European Union's Committee on Culture, Science and Education proposed to the Parliamentary Assembly of the Council of Europe that 'creationism' (a carelessly-defined, pejoratively-used word) should be outlawed. It is as if such bureaucracy, examining the intelligent design of a motor car, banned mention of its inventor from science classes in its federation's schools! How could you argue from interpretation of the facts for an inventor when, by philosophical decree, you must desist?

Suppression's next step is imprisonment and, indeed, in Marxist China members of the Communist Party may be punished for questioning the creed of scientific atheism.

British governmental creed may well be seriously infected too. Scientism[19] is not science. It is a world-view dictating that only materialistic answers to any question whatsoever be considered. Science and the Soul demonstrates the nonsense of such philosophical *diktat* which is, essentially, an affront to scientific free-thinking. This is because, as an extreme that *excludes* open-minded holism, it accedes to an imposition of materialistic interpretation of data concerning historical origins. 'Forget an immaterial element of information; this does not exist!' it cries. *Thus legal imposition due to pressure from a fashionable lobby would reduce abductive scientific query[20] to atheistic consideration*

---

[19] see Glossary.

[20] see Glossary: logic.

alone; and thereby imply that this alone is scientifically/ naturalistically correct! By now you see Design is not a pretty theory all alone. All sorts of people have designs on it - some nice and others not so nice.

There is no need for all this statutory nonsense. In the application of abductive reason frame of inference, the mind-set of interpreter of fact, is crucial. The mind-sets here *inclusively* involved are basically two - materialistic (scientific) and holistic (scientific plus). What does the 'plus' imply? 'Plus' includes, besides material, immaterial elements. Thus its anathema allows and not excludes purpose, teleology and design as factors in a valid explanation of how natural systems have emerged and, therefore, how the world works.

Which truth of our origins is rational, which irrational? Which true, which false? *Has, for example, physical energy (Darwin called it natural force) the power to generate a living body? Can you shoe-horn every finding into evolution's box?* **The fact is all machines work using natural forces but these do not explain their origins. Each machine embodies the purpose of its maker.** And what are bio-bodies but the most amazing soft machines? They incorporate signals, receptors, many kinds of mechanism, coordination, integration, regulatory components and factors regulated to produce intended ends. They can even reproduce themselves! And the only known cause of such factors and their mechanistic functions is, as all engineers understand, immaterial and called intelligence. Not only is any specific body structurally coordinated and homeostatically operative but it is *codified* to be so! Symbolic plan is psychological; mind *anticipates* an outcome that blind forces *never* could. No more is such foresight as implicit and yet obvious as in the reproductive development of life-forms. Maybe, you counter, 'design' must be correctly read as Lady Luck's top gift; but, equally and much more cogently, it may be argued that it's not. **Looking to a future target never was the gift of chance.** Intelligence is perhaps, though heterodox, better value as an explanation of informed biology than orthodox Darwinian currency.

Thus, from this perspective, simply know that chasing after ape-men might turn out an exercise that's, from its first step, philosophically flawed.

Nevertheless, as we have just observed, there are governments (not only communistic ones) whose lawyers seek to legislate promotion of one pseudoscientific view at the expense of its antithesis. *Would an imperative of law, whose basis is fair justice, leave only the materialistic arm of dialectic unrestrained? Would it police imbalanced skew?* **Since where we came from radically affects our frame of thought the pressure of such biased legislation must affect a subjugated population's mental health.**

7. Racism, Slavery and All That

Survival of the fittest by natural selection implies superiority of the survivor. Darwin's theory proposes the gradual, unpredictable evolution of species fitter in survival stakes than their forebears. Thus, from ape to man, inferiority logically improves through ape-men, archaic humans and contemporary savages to, for a Victorian, white dominance. **There's no doubt the theory is intrinsically racist.** It is not evil of itself but its internal, eternal driver - warfaring competition to survive - may easily be used as an excuse to justify, condone or tolerate those actions labelled, morally, as crimes. After all, what morality is there inherent in material oblivion or, as a consequence, materialistic world-view?

Humans are, genetically, perhaps 99.9% identical. If they descended from an original pair they would obviously all be, in this sense, blood relatives. Brothers and sisters. However, the title of Darwin's landmark tome is 'The Origin of Species by Means of Natural Selection or The Preservation of Favoured Races in the Struggle for Life'. **Clearly, in this context 'race' approximates to 'species' but, where 'race' is nuanced with an innate sense of superiority-and-inferiority, Darwin was, with the great majority of his contemporaries, of a supremacist, racist mind-set.** Did his avidly-supported half-cousin Francis Galton's pseudoscientific study of eugenics not imply such rationalistic moral code as 'fittest races' might inflict on cavemen, primitives and those in poor or technologically backward, 'third world' states? It could be argued that twentieth century materialistic shades of social Darwinism (as diverse as totalitarian fascism and communism, socialism, colonialism, racial supremacy, dog-eat-dog capitalism, scientific atheism, hedonism, humanism, anarchism and sheer nihilism) have all, in some cases at the cost of vast human depression, inhuman suppression and cruel suffering, drawn doctrinal strength from the material heart and pumping blood of evolution theory's 'unofficial bible', The Origin of Species.

The Fuegian Indians that Darwin encountered on his voyages were primates but not apes. They were a people who lived, naked, among the rocks of ice, wind and fire in Tierra del Fuego, at the southern tip of South America. They numbered about 7000 in Darwin's time. By 1932 there were only forty-two survivors, including half-breeds, and now they have probably disappeared. Darwin's opinion of them gave his contemporaries a comfortable feeling of their own supremacy.

"I could not have believed how wide was the difference between savage and civilized man; it is greater than between wild and domesticated animals, inasmuch as in man there is a greater power of improvement... the difference between a Tierra del Fuegian and a European is greater than between a Tierra del Fuegian and a beast." [21]

Three Fuegians were captured and brought to England for education. They stayed for a few years and were returned. Darwin talked with them on board ship. He found them quite human in their sympathies but scarcely articulate. Captain Cook compared their language to a man clearing his throat and Darwin came to the conclusion that it consisted of only about a hundred different sounds. Many animals make a dozen or more different sounds so that comparison with advanced animals seemed apt enough.

Of course, man's body is animal but he has gifts which distinguish him. Certainly, in later life, Charles Darwin was so astonished at the apparent 'rehabilitation' wrought by missionaries on natives that he contributed regularly to the funds of a missionary society. But, although Christianity might help the most degraded savages evolve into more manlike specimens, it did not change his philosophy about them. In 1921-3, however, when the traditions of the tribe were still intact, two Austrian priests who had been trained in anthropology went to live among the Fuegians. They found the tribesmen set a high standard of morality and ethics; they believed in a Supreme Being who had created the world and the framework of society and prayed to Him in need, especially at death. An English missionary, Thomas Bridges, made Tierra de! Fuego his home and brought his family up there. He found the natives moral, kind and sociable. They had respect for family life and were not cannibals. Mr. Bridges spoke the language of the tribe. He compiled a dictionary that was not exhaustive but contained 32,000 words and inflexions. The vocabulary was rich and the grammatical constructions involved. This was the language they spoke from an early age, and throughout life.

Darwin was wrong about the Fuegians. Myopic, due to the imposition of his own world-view, or blinkered because he could not communicate with them, he misjudged them. They were true men as much as he was, unschooled but with full intellectual faculties and spiritual qualities that they did not show to casual passers-by. If Darwin could misjudge a living 'specimen' with whom he could communicate, what trammels are there to restrain the speculation of an ardent evolutionist (and his artist associate) over a few broken bones?

Theocratic hypocrites may pervert religion to support various kinds of savagery; but Darwinism, centrepiece of liberal, secular philosophy, *demands* progress by way of superior competitors. Lower varieties represented part of a ladder of 'inferiority' reaching up to 'us'! Such

---

[21]   Barclay V. *Darwin is not for Children*, Herbert Jenkins, 1950, Chapter 14.

concept is embodied in Clark Howell's outrageously deceptive but iconic, textbook evolutionary sequence referred to in *Cartoons*. Colonialism's Darwinists saw their humanistic rationality, technology and education as far above the state of their subjects. Such evolutionary attitudes did not, it has been said, *cause* exploitation and slavery (which long preceded Darwin and succeed him still) but rather serve to justify. **How, if nature lacks morality, can such behaviour be 'wrong'?** It's natural!

Blame God not Darwin! Survival and control! A violent streak in centaur man[22] has always sought, in the interest of selfish, physical survival, to differentiate and dominate. Whosoever's not in my gang is to be attacked, abused or, in the extreme, exterminated. When, in all history, have not bullies kicked, the weaker suffered and their slavers thrived? When has politics, in order to expend with or exploit, not demonized 'those others'? And theory strains to segregate Neanderthals although, to prove they were inhuman, you would have to try and interbreed. This is not possible but still the British proved sub-man's humanity. They committed shameful genocide of which remains survive. Between 1804 and 1876 the local Tasmanian population was entirely wiped out. Of course, before this happened slavery and sexual relations generated mixed blood descendants, thousands of whom survive. Nevertheless, 'scientific racism' of the day held that Tasmanian aborigines were a 'missing link' between stone-age primitives and modern man. Erasmus Darwin had an aboriginal dug from the grave to initiate a large collection of stuffed Tasmanian exhibits at The Royal College of Surgeons.

His grandson Charles, whether or not implicated in the Tasmanian terror, wrote that the difference between a primitive from Tierra del Fuego and a European was greater than between the native and a beast (see also Chapter 26: A Continual Fight). And Charles' relation, Francis Galton, used the measurement of skulls (discredited phrenology) to gauge human status; he also initiated the study of eugenics, selective breeding that might, in the manner of horses, dogs or sugar beet, drive human evolution, by the abortion/ post-natal elimination of inferior/ unwanted stock, into superhuman gear. Does such a notion figure as an undercurrent in today's genetic engineering? If so, beware the tendencies!

In the 1890's Friedrich Ratzel had intellectualised the idea of *lebensraum* (living space) - one already practised with vigour by the systematic, military massacre of black locals in German West Africa. Then in 1904 Alfred Ploetz, founder of the German Society for Race Hygiene, wrote to Galton praising his work. Later Heinrich Himmler

---

[22] see Glossary.

praised Ploetz. Galton was therefore one deft step removed from the intellectual responsibility for Hitler's death camps. A direct consequence of the holocaust was the establishment of a home for the Jews, Israeli-Arab grievance and the religious veneer of global terrorism. Doesn't Galton's ghost haunt you as well? Did he learn nothing from 'progressive' Uncle Charles? Not only did an evolutionary train of thought justify and thereby foster attitudes of racism but also various imperial malpractices and 'medical' experimentation with convicts. It also (Chapter 26: The Nature of Evil) raced directly to justify and thereby bolster fascist, communist and other political forms of tyranny, murder and genocide. The philosophers and scientists who developed neo-Darwinism were eugenicists but, post-holocaust, public discussion of this issue was abruptly shelved.

Was the idea dead? What about promotion (or enforcement?) of social and biological improvement (= evolution)? In the 21st century might Richard Dawkins or Peter Singer venture to ask whether eugenic selection is all that bad when it comes to selecting for musical, mathematical or athletic talent and, certainly, promoting a vigorous population and culling the disabled? *As the basis of secular thought, evolutionary progress and, if necessary, evolution helped along by man is so engrained in the modern mind that it now subliminally bolsters attitudes towards all manner of controversial issues ranging from abortion clinics through cloning to embryonic stem cell research.* Aren't prenatal selection tests and death camps for the unborn now in legal operation? Aren't chambers of that holocaust in action in the wards of hospitals that, on demand, exterminate unwanted, inconvenient, healthy but still unborn babies by the million? Such sanctity! A brave new world of 1984 might well, on such a slippery slope, slip in. But never mind. Someone will tell you how he's reasoned it's alright.

Why aren't apes turning into humans now; or humans into superhuman species? The evolutionary pursuit of 'living links' has also led to sad, individual cases of cruelty. For example, the Belgian colonial authorities exported many pygmies (a type already habituated to enslavement by local tribes) to European and American zoos. One such Congolese primitive, Ota Benga, was in 1904 torn from his family, taken to the USA and, among other exhibitions, caged with various apes in the Bronx Zoo. The zoo's director, learned, expert Dr. Hornaday, was in tune with prevailing sentiment. He waxed lyrical about his 'transitional specimen' until, unable to bear the treatment any longer, 'it' committed suicide. Nor was Ota the only one. Since a 'racist' attitude to 'inferior species of native' was the cultural norm in Darwin's day, check for yourself the unsurprising suffering of eskimo Abraham Ulrikab's family in Europe.

Poor Ota Benga wasn't even 'real' but why not mint a million *real* ape-men? Is this the currency of evolution's great idea, 'survival of the fittest'? Is this where black philosophy can lead? It was certainly passionate atheist and evolutionist Stalin's direction. In 1936 he rejected a eugenic proposal to create (in the Nazi style) a superior Soviet population by inseminating women with sperm from a few 'top quality' donors. In 1926, however, the Academy of Sciences was ordered to create ape-men. All other primates are considerably stronger than man (a significant reversion in the survival stakes if we evolved). A hybrid with chimp intelligence and human strength would make a poor survivor but one with human intelligence and chimp strength could be dangerous. So was born the mad idea to mimic by unnatural selection a process, hybridisation, which evolution theory insists has advanced organisms many times. This time it would force-breed an army of pain-resistant, invincible humans - living war machines' called humanzees. And, at the same time, prove evolution and discredit the hateful but still powerful church. Whatever the motive biologist Ilya Ivanovich Ivanov, whose scientific pedigree was in the line of artificial insemination and who had already considered the possibility of creating a hybrid ape-man, was engaged to execute the task. Man as Creator!

In Conakry, French Guinea, Africa, the promethean upstart tried to cross human sperm with female apes and failed. No doubt (see Chapter 21) hybridisation between primates such as gorilla/ bonobo and baboon/ macaque can occur in the wild. Sterile specimens have languished in zoos. And there were rumours of *Gigantopithecus*,[23] almasty and yeti. The experiment had not worked but why shouldn't it? So in Sukhumi, Ivanov retried using orang-utan sperm on human females. No-one knows if any died but no hybridisation occurred. Ivanov was subsequently 'purged' and exiled to Kazakhstan where, in 1932, he died while still working at the Kazakh Veterinary-Zoology Institute.

But it is unlikely he will be the last of it. Some fervent biologists now claim they can take evolution over. They can mix and match genetic profiles empirically finding, according to commercial orders, what works best. Mindless nature needed luck but, using biotechnology, they want to create on purpose. Unmindful of the nemesis, what 'eugenic' monsters will they dare concoct, what demi-organisms (called chimeras) will they arrogate to make? Perpetrators, even criminals, can always find a reason;

---

[23] *Gigantopithecus blacki* was first identified in 1935 by von Koenigswald in one of his habitual trawls of Chinese apothecary shops looking for unusual teeth. He considered a large, now-missing molar 'human' and classified it in honour of Davidson Black. Three or four mandibles (lower jaws) and more teeth but no other bones have excited much speculation (not least by Weidenreich) and the classification of a couple more extinct species.

but, mindful every powerful technology is also malleable to evil purposes, is genetic mix 'n' match a game that wise men should embark upon? The morality of evolution is 'survival of the fittest'; on this raw ground could you defend transgenic engineering's ethical or moral case? Or is morality a hyper-reasonable function of the sanctity of well-created forms of life? Materialism scoffs at such 'creationist' mentality; scientific atheism scorns it as an opiate, uncurious, 'recessive' and a mutant weakling's code. By which mind-set, though, is rightness found or human truth received?

In summary, you cannot blame a concept for the sins of man. Darwinism cannot generate them (though, if true, long evolution might have). You *can*, however, deprecate the fact that it intrinsically justifies them. How can a theory based on 'survival of the fittest' whole-heartedly and logically condemn wars for supremacy, exploitative enslavement of 'inferiors', genocide of lower races, eugenics, abortion or liquidation of the sick and 'lesser' humans in our race? Racism was culturally acceptable only a few decades ago but we shall see later how its current, politically correct excoriation has influenced the theory of evolution considerably. Just as a single skull may inspire different images, so the theory itself is flexed for different reasons.

8. *An Instructive Analogy*

So far this book has concentrated on the will to evolutionary interpretation of facts, on the irrationality inherent in evolutionary announcements, on plausible but not necessarily true arguments proposed in support of materialistic theory; also on skew, imperative and the creation, due to wishful thinking, of self-deception, fraud and pseudoscience. *One particular febrile Son of Evolution demonstrates these characteristics in quantity. He affords an instructive analogy that more than summarises all that's been discussed and affects (or infects, depending on your viewpoint) what follows.*

As a professor at the University of Jena from 1862-1909 Ernst Haeckel's ideas held sway. Of three main strands in his thought - Romanticism, Materialism and Darwinism - only the latter concerns us. As a scientist Ernst could well use a ruler, microscope and reference book. He was also a field researcher and draughtsman of exceptional talent who discovered, named and described thousands of new species (mostly radiolarian and medusae). To these he added the invention or popularisation of a number of evolution-skewed terms (such as phylogeny, anthropogeny, heterochrony and polygenism) and fresh hypotheses (such as his Theory of Recapitulation).

He was, moreover, an avid, self-appointed propagandist for Darwinism in Germany, though his imaginative 'Darwinismus' expounded views far removed from the biological ideas or philosophical views of Darwin himself. He believed in, for example, Lamarck's non-Darwinian mechanism rather than natural selection and that man emerged from Asia.

English Darwinism interlinked two main themes, natural selection and the struggle for existence. Social Darwinism is an attempt to explain human society in terms of evolution but Haeckel's interpretation was quite different from that of capitalist Herbert Spencer or of communist Marx. He professed a belief in a literal transfer of the laws of biology to the social realm. For him a major component was the ethic of inherent struggle between higher and lower cultures, between races of men. There also existed a rejection of rationalism; and with it the embrace of *Blut und boden* (blood and land), man and nature, rootedness with earth. The German form of social Darwinism became a pseudo-scientific 'religion' of nature-worship combined with notions of racism. This flared up into a full-bodied ideology with trends of imperialism, romanticism, anti-semitism and, as a racially united and powerful Germany, nationalism. As a reputed scientist he lent Volkism a respectable and appealing image but the movement was proto-Nazi in character; and the Monist League (established 1906) laid, with its embrace of evolutionary biology, the ideological foundations of National Socialism. As well as Nazism the mode of thought supported Samuel Morton's 'scientific racism' which was used as a confederate argument to justify abovementioned slavery. Josiah Nott even composed a series of diagrams purporting to show how negroes were cast between 'a Greek and a chimpanzee'!

Be that as it may, in 1866 Haeckel the apostolic missionary met his hero, Darwin, at the latter's home (Downe House in Kent). With 'bulldog' Huxley he pioneered the notion of serial arrangements of data as 'proof' of evolution. And, shoulder to the wheel of building evolutionary momentum, he plausibly presented and proselytised the notion of 'living matter' that could, of course, self-organise into organisms; [24] he drew various trees of life; hypothesized, described and drew several proto-organisms which have not been found to exist; generated the theory of recapitulation whose incorrect catch-phrase, 'ontogeny recapitulates phylogeny' is still shamelessly recycled in biology textbooks; played a role in the ape-man saga; and in the development of social Darwinism with its element of 'scientific racism'. In a nutshell, Haeckel's diverse and fervent threads of evolutionary construction have permeated western thinking since. Indeed, for holists he created a plethora of mischief everywhere!

Firstly, Haeckel's mystical belief in the forces of nature as 'living matter' is a wishful fantasy that need not concern us here. For Trees of Life see *SAS* Chapter 22 and Wells 'Icons of Evolution'.

Next let us, as far as the origin of life is concerned, go back to 1860. Pasteur, a creationist, concluded experiments a year after the publication of *The Origin of Species,* showing that broth in sterile flasks did not spoil. Microbes from outside could not get in, and germs could not generate spontaneously inside the flasks although both broth and air inside were ready to support them. His work served to demolish the primitive notion of 'spontaneous generation', held at least since the time of the Greeks, and supported the law of

---

[24]   refer, for a book which falsifies ten 'icons of evolution' still reiterated in schools worldwide as 'proofs' of evolution, to 'Icons of Evolution' by Jonathan Wells. These icons are The Miller-Urey experiment, Darwin's Tree of Life, Haeckel's embryos, homology (e.g. the pentadactyl hand) in vertebrates, *Archaeopteryx*, peppered moths, Galapagos finches, fruit fly experiments, fossil horses and, directly involving us, ape modified into man.

biogenesis which holds that life comes only from living material of the same kind. 'Only from a previous cell comes another cell'. This law throws no light on the origin of life-forms, although they must, like the universe itself, have had a beginning.

Not everyone was happy with Pasteur's results, or with their implications for life's origins. If spontaneous generation did not occur then life came to earth either from space (and how did it originate there?) or by an act of supernatural creation (excluded by definition from scientific investigation). Therefore, a materialistic explanation of origins had to be revived. However, applying artificial respiration to the dust of the earth puts a scientist in the awkward position of having to abandon a basic law - biogenesis - that has been universally validated and has no known exception. To repeat, biogenesis states that a living thing can originate only from a parent or parents essentially the same as itself. Huxley, Haeckel and Darwin found themselves in this fix in the mid-nineteenth century and 'solved' it in various ways. The former couple seized at the myth of *Bathybius,* while Darwin dreamt of a prebiotic pond that doubled for the Fountain of Life.

At that time protoplasm was all the rage. Many a scientist was happy to agree that he could trace his ancestry back to a protoplasmic primordial globule of slime. The gap between living and non-living was one which the Darwinian evolutionist was obliged to fill. Haeckel, almost completely ignorant of cell chemistry and genetic operations, proposed a hypothetical precursor to the amoeba, the moneran, 'an entirely homogeneous and structureless substance, a living particle of albumin, capable of nourishment and reproduction'.

The search was on. In 1868 Huxley, examining mud samples dredged off north-west Ireland some ten years earlier, identified a jelly in which were embedded tiny calcareous discs (coccoliths) which he incautiously linked with Haeckel's moneran. In honour of Haeckel he called it *Bathybius* ('life of the deep') *haeckelii* and, speaking before the Royal Geographical Society in 1870, maintained that it formed a living scum on the sea bed extending over thousands of square miles. There was great expectation in 1872 as HMS *Challenger* steamed out of port on an expedition to explore the world's oceans. However, no more *Bathybius* was found. Indeed, Mr. Buchanan, the expedition's chemist, observed that he could produce the characters of the indescribable animal simply by adding strong alcohol (such as was employed to preserve biological specimens) to mud. A specimen examined under a lens showed that calcium sulphate was precipitated in the form of a gelatinous ooze which clung around particles as though ingesting them, thus lending superficial protoplasmic appearance to the solution. Thomas Huxley's sample had been thus contaminated. Although it lingered in Haeckel's mind, for everyone else *Bathybius* died the death.

But theory demands abiogenesis be true and thus the search for cells arising out of chemicals[25] continues unabated to this day. **Thus not atomic but a mythological vitality evolves**.

Do you remember Howell's 'March of Progress'? Is progress the same as development? *What has biological development of man, from egg to adult form, to do with evolution of an ape perchance to man?* **The answer's 'absolutely nothing'. Development is programmed therefore planned. It is directed to an end. But, in evolution, there is nothing preordained and, therefore, rationally explicable.** In fact, if one coincidence in all the train alleged to run from ape to man had ricocheted another way you'd not be here. You are unarguably a conceptionless inception of Immaculate, Pure Lady Luck. **How then was codified developmental process subverted to the yoke of evolutionary propaganda?**

Drawings by Ernst Haeckel (1874)

| fish | salamander | turtle | chicken | rabbit | human |

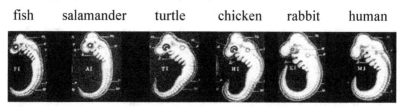

Photographs by Dr. Michael Richardson (1997)

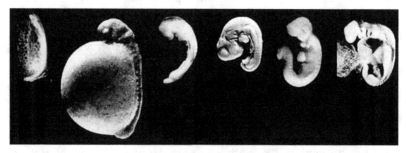

8.1 Developmental Misrepresentation by Haeckel.

In 1828 Karl von Baer, the founder of embryology, published hypotheses based on observations of the similarity of embryos of mammals, birds, lizards and snakes during their earlier stages. This makes sculptural sense. When a sculptor starts to carve a block of marble

---

[25] see also *SAS* Chapters 20 and 21.

into a statue, he immediately reduces its potential. Broad outlines are established and, eventually, distinct details of his unmistakable work of art emerge. This is exactly the way development works to reduce generalised, shapeless genomic potential into a *pre-defined* conclusion. His correct assertion that 'generality of embryo precedes increasing specificity of later stages' refuted earlier ideas of 'preformation' (e.g. that a sperm was a homunculus or man in microscopic miniature). In addition, non-evolutionist von Baer noted that the embryos of animals never resemble any other adult animal, only other embryos.

However, in 1868 the deliberations of Haeckel stepped in. After all, in 1859 Darwin had, adopting von Baer's 'generality' in a way it was not intended, suggested that embryology might provide a mechanism for evolution due to 'the fact' that embryos of species widely dissimilar in adulthood show 'similarities'. He imaginatively construed that such similarities demonstrate descent from a common ancestor as well as that ancestor's adult appearance! *Thence Ernst converted the work of von Baer according into his own convention of a 'fundamental biogenetic law'. It was encapsulated in the phrase 'ontogeny recapitulates phylogeny'.* This, Haeckel's reinterpretation of the law of biogenesis, was supposed to mean that a vertebrate, in the course of its embryonic development (ontogeny), successively passes through (recapitulates) phylogenetic (or evolutionary) stages evolved through by its ancestors. Although the theory does not apply to the plant, fungal or bacterial kingdoms nevertheless Darwin himself saw it as fine support for his theory. And successive textbook repetition of its selective 'evidence of evolution' has meant that Ernst's push has propelled generations of students, including some today, up to evolutionary speed - although, however hard it dies, his interpretation has long been discredited. How?

Von Baer's work was based on careful observation but Haeckel's on theoretical deduction from evolutionary theory. **To support his deduction, an 'icon of evolution', Haeckel faked evidence by both omission and commission of facts.** By 1874 his tweaks had been declared invalid by Professor His. When charged and convicted of fraud by a university court at Jena, he agreed that a small percentage of his drawings were forgeries. He was, he pleaded, merely filling in and reconstructing missing links where the evidence was thin. *In fact his main faults had lain not only in disingenuous copying (he used the same woodblock to demonstrate similarity between different classes!) but also in the selectivity of his evidence.* The avowed intention was, of course, to demonstrate that early embryonic similarity showed common ancestry. He did not, however, start at the zygote where different classes of egg differ greatly in yolk content, size shape and in cleavage patterns and the organisation that prepares them for gastrulation. These primary stages actually refute his and, therefore, the Darwinian case. So Haeckel made his first stage (see picture) one where, post-neurula, a superficial appearance of convergence may seem to appear - in other words when diverse early stages briefly and superficially converge before subsequent divergence into organ and body formation. Even then they are actually substantially different in size and shape but Haeckel minimised this fact. For example he omitted 'anomalies' and drew embryos in the 'tail-bud' stage of development as almost identical. They are not. Moreover mammals and frogs have limb buds at this stage but for the sake of 'evolutionary conformity' Haeckel 'air-brushed' them. Professor His accused; Adam Sedgwick (Darwin's tutor, lived 1785-1873) clearly argued that early similarity and later dissimilarity was not in accordance with the facts of development. He rejected the notion of 'ancestral rudiments' comprising developmental steps. Thus from Haeckel's contemporaries to present-day professional embryologists disagreement and accusations of fraud, self-deceptive misrepresentation or plain, theory-driven error have accumulated. Again, the mid-20th century, Sir Gavin de Beer flatly refuted 'recapitulation'.

We can elaborate. The spark of fertilisation trips cleavage, gastrulation and neurulation then, after these stages have occurred, Haeckel's first stage is obtained. *These 'pre-Haeckelian' stages should, by his logic, be even more similar to a primordial ancestor. But they are not, any more than the highly codified potential of a zygote represents the first recapitulatory organism.* In Ernst's upside-down world a single-celled zygote represented our primal ancestor but he omitted to mention that multi-cellular organisms actually differ considerably with respect to eggs/ seeds; and he ignored the following, mechanically essential 2, 4, 8 and 16-celled embryonic stages because there is no known organism, extant or extinct, whose adult body is (due to the facts of surface tension) composed of these quantities! He also left out the different ways by which those 2, 4, 8 and 16-unit stages are expressed. For example, mathematician D'Arcy Thompson noted that twelve patterns of segmentation were possible at the 8-piece stage. In fact, initial 'sketches' that lead from a solid ball of cells (the morula) to the formation of a hollow, 32-celled ball (the blastula) are different in amphibians, reptiles and mammals. Hence developmental courses diverge. They may vary even in closely similar forms so that apparent homologies may not, from their different developmental pathways, actually be so. Whichever way, such courses lead every embryo to expedite the features of its genus. For example, human definition is increasingly apparent from its glimmerings after about a month to clarity after six weeks.

A comparison of photographed embryos set against Haeckel's drawings made by embryologist Michael Richardson (Anatomy and Embryology Vol. 196 (2), ps. 91-106, 1997) clarifies the scale of his deception. Even the confession at Jena was, by this measure, a fraudulent diminution of this scale. Although the theory of recapitulation has been decisively abandoned by science for well over 100 years the will to believe despite the facts cannot be overlooked. In 1901 the disgraced set of drawings was placed, gill arches and all, into an educational volume and has, from that time, been parroted by less than perspicacious authors of high school textbooks. *And if the textbook authors were plain ignorant then knowledgeable experts, such as Stephen Gould, kept quiet in order not to publicly give the game away. In fact, both accessories to the offence turned on 'creationist' whistle-blowers for the real offence - anti-Darwinism!*

In these matters Haeckel only saw what he was looking for. He then transformed his skewed observations according to the imperative of Darwinian logic. Finally, he employed his constructions and reconstructions with missionary zeal to make the world believe that evolutionary materialism is the truth. Never mind that his embryological drawings were 'incarnations of concepts masquerading as neutral descriptions of nature'. Never mind that you know a devil by the quality of his deeds. How great, rather, is the power of propaganda. Darwin rewrote Baer's observations and ideas. Haeckel amplified the issue. *Some biology teachers still believe and teach his mumbo-jumbo! Is it with this sort of tactic (wherein wishful agendum trumps reality, distortion of the truth wins victory and the Darwinian mythology's recycled) with which the ape-man saga's been promoted?* **As regards imperative and skew is the Haeckelian analogy instructive?** We shall see.

9. When Is a Man a Man?

When is a man a man, an ape an ape and either not the other?

What is the smallest category into which similar individuals can be grouped? Biology needs the concept of species, or something like it, to identify what it is talking about. Aristotle called such particular form or kind εἶδος.[26] Cicero, in turn, used the Latin word for εἶδος, species. Unsurprisingly, in his fourth-century translation of the Bible into Latin version called the Vulgate Jerome uses the word 'species' to translate the Hebrew 'min'. In the later King James English version God created every living plant and animal ('whose seed is in itself') from the earth 'after its kind'. *What is 'kind'?* Is it each one's archetypal idea according to whose pattern earthly chemicals were shaped? Does the use of word also imply that organisms only reproduce after their own type? *Again, what if any is the limitation of such type?* Can it, as the theory of evolution needs, be limitlessly extrapolated such that a proto-cell could father all the living world? Or, as the Law of Heredity indicates, are parents observed, without exception, to reproduce their own kind - within which plasticity is limited.[27]

Carl Linnaeus, the founder of modern taxonomy, was a creationist. The academic language of his time was Latin. Thus, probably following Cicero's use of the word, he counted 'as many species as there were different forms created in the beginning'. Or, unlike Cicero, did he mean species in the modern, scientific but non-Latin sense such that every organism so designated by a taxonomist was separately and invariably created. If so, there is debate over whether he later repented of such error - for both modern evolutionists and design theorists recognise it as such. Nevertheless, no less than the Concise Oxford Dictionary *still* peddles this error by defining *speci*al creation as 'the belief that every species was individually created by God in the form in which it exists today and is not capable of undergoing any change'. *If the book means 'biological species' we can see how such patent nonsense arose by misunderstanding and unfortunate translations; and also that, since **no-one educated believes it**, it should now be deleted.*

In short, the level 'species' may certainly not be the same as 'type' or 'particular kind' but, equally, placing organisms into the taxa of classification contrived for human convenience may not entirely reflect the natural world either. There are, for example, splitters and lumpers. Splitters are taxonomists who tend to label a specimen a 'new species' on account of minor morphological difference. Less fastidiously, lumpers tend to a broader approach citing the criterion of interfertility to label either a species

---

[26] see *SAS* Chapter 16: Platonic ἰδέα meaning form, species or archetype.

[27] *SAS* Chapter 22: Origin of Species and Origin of Type.

or sub-species. Both cannot be right. Such split-or-lump definitions, as that of species itself, are ones of convenience but can cause confusion not least, as we shall see aplenty, in the naming of human fossils.

Darwin preferred, as opposed to the Linnaean interpretation, to 'plasticise' the idea to what he saw as its cause - gradual change. He wrote "*I look at the term species as one arbitrarily given, for the sake of convenience, to a set of individuals closely resembling each other, and that it does not essentially differ from the term variety.*"

A species is therefore, in the blurred dynamic of evolutionary time, a 'variety'; and variety incipient new species. *In fact, nobody has a problem with an origin of species - although taxonomists still argue over what the label actually means.* Ring-species, for example in the case of some gulls or sparrows, may interbreed in linked populations but, where distantly related end-populations 'close' the ring, do not. Thus two species have geographically emerged from one. **Call the process micro-evolution if you want; but Darwinian-style variation is not evolution (even if you call it micro-evolution).** Gulls stay gulls and sparrows sparrows. No way does this amount to macro-evolution, that is, evolution of fresh body-forms.

9.1   Unlimited Plasticity

The 'progressive' view of speciation involves gradual steps the order of which Darwinists try to compile in series of more or less similar fossils. This is called, in *SAS*, 'unlimited plasticity'. Both to amplify the conceptual idea and for visual impact bone fragments may be fleshed out using computer graphics or photographs of models.[28]

However, Darwin not only regarded varieties as incipient species but also proceeded to *extrapolate* on the hypothetical principle of unlimited plasticity. He *did* consider it a minor stage in the long process of macro-evolution. Such extreme extrapolation, variation-*without*-theme, formed the very basis of his theory. Thus, because natural selection[29] works on these varieties, it would seem that the origin of species is what, in its flexible gradualism, in the end denies them any durability. Species are simply moments in flux: snapshots of change. The iconic model of speciation[30] derives from his observation of thirteen or so finches on the remote Galapagos Islands. It illustrates how, by various mechanisms, one group of descendants from a single population may diversify and perhaps cease to interbreed. The question

---

[28]   e.g. Alice Roberts (2011): Evolution The Human Story.
[29]   see *SAS* Chapter 22: The Editor.
[30]   *SAS* Chapter 22: Galapagos and All That; also *fig.* 22.1.

is how far such variance extends. Is it simply elasticity within an archetypal theme or can it be *extrapolated* past macro-boundary to the point where, say, a mouse snaps out of 'mouseness' into something else? On the principle of unlimited plasticity does speciation accumulate so that this stretch and snap of boundaries keeps happening until the mouse is made a whale? Or has been transformed through ape to man? Is the integrity of archetypal 'rings' paramount or can 'broad' *micro-evolution*, against all experimental evidence, proceed unchecked into *macro-evolution*? Can its unlimited plasticity transform bacteria into humans? This is the dogma. It is precisely why Darwin's disciples recoil at the restrictive anathema of 'archetype'.

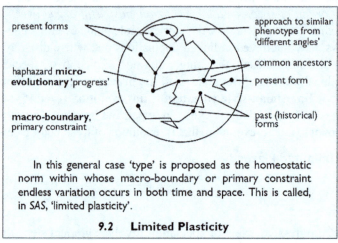

In this general case 'type' is proposed as the homeostatic norm within whose macro-boundary or primary constraint endless variation occurs in both time and space. This is called, in *SAS*, 'limited plasticity'.

9.2   Limited Plasticity

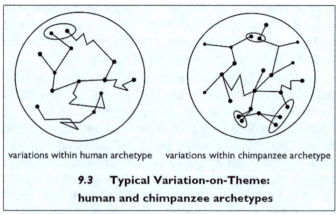

variations within human archetype   variations within chimpanzee archetype

9.3   Typical Variation-on-Theme:

human and chimpanzee archetypes

Similarly, in this view the circle or primary constraint represents εἶδος, species (in the Latin sense) or archetype within which circumstantial, so-called 'micro-evolutionary' variations occur.[31] In this case two types are conceptualised.

---

[31]   biological variation is discussed throughout *SAS* Book 3.

Variation-unto-theme; small variation writ large is the Victorian sage's gamble and, if in the real world macro-theme runs unto micro-variation, his great, anti-parallel mistake. *Top-down,* on a genetically demonstrable principle of limited plasticity, typical variation might be thought of as different permutations of a single genetic potential expressed at different times and places. Genetic potential constitutes the amount of variation that a type of organism can produce from operationally-organised genetic material already present and acting as a 'variation bank'. Such adaptive potential (see Glossary) is embedded in the typical genome; its intrinsic, preordained flexibility is realised by switches such as sexual recombination, *RNA* factors, epigenetic markers, environmentally-sensitive proteins and so on. *In this computative way the definition of speciation might be seen as differentiation circling round a norm, wild-type or archetype.* <u>*In other words, the reverse of Darwin's 'sagacious' gamble.*</u> *Theme-unto-variation; variation-on-theme.*

Your interpretation is a function of your starting-point. **Is it chance or information?** This central point demands recapitulation.[32]

Bottom-up, evolutionary biology assumes that, due to *unlimited plasticity* generated unpredictably by chance *c*ontinual variation (~micro-evolution) leads to macro-evolutionary transformism. Thus natural laboratories of evolution, such as isolating conditions, rapidly yield (by 'founder effect' and subsequent 'adaptive radiation') myriad examples of 'allopatric speciation'. These include species of finch, honeycreeper, fruit fly and so on. Islands such as Hawaii, Madeira, Madagascar and the Galapagos are cited as Darwinian boilerhouses.

Top-down, however, interprets the same effects in terms of *limited plasticity,* that is, continual variation within the powerful adaptive potential of an archetypal program. Of course, through time genetic recombination and epigenetic response to environmental conditions demonstrate the intrinsic flexibility of this informative potential. *Why should man alone be immune from the nature of adaptive radiation from an original archetype and a consequent series of 'founder' populations?*

Accordingly, proof-reading of *DNA* copy ensures that actual text (carrying in-built, codified adaptive potential) is printed with as high-fidelity a quality as possible. *Uncontrolled variation by mutation is resisted.* **And, when genetics decisively demonstrates the elastic limits past which mutation does not survive, it also demonstrates conclusively, through myriad experiments (e.g. with *E. coli* bacteria, fruit flies and mice), that the macro-evolutionary principle of unlimited plasticity (or unbridled extrapolation) is incorrect.** Ask a

---

[32] see also *SAS* Chap. 22: 'The Origin of Species' and 'Galapagos and All That'; also Chap.23: 'Evolution in Action?' and 'Supercodes and Adaptive Potential'.

dog-breeder if he has bred other than a dog. Ask him why. Why, again, should the situation differ for mankind?

Charles saw the problem - even if you loan it bags of time. "The geological record is extremely imperfect and this fact will to a large extent explain why we do not find interminable varieties, connecting together all the extinct and existing forms of life by the finest graduated steps. He who rejects these views on the nature of the geological record will rightly reject my whole theory." **Where Darwin staked his claim on continuity,** *discontinuity* **(the primary prediction of an archetypal model) is still what we actually find**. Nobody is more aware than a palaeontologist that Darwin's 'interminable varieties' are missing from life's petrified family album.

Now we reach the modern but, in practice, occasionally blurred taxonomic definition of species - a population of organisms reproductively isolated (by various factors including Huxley's criterion of geographical isolation) from a similar group. In other words, members of the same species can interbreed with each other but not cross-breed with other species. This definition is clearly vague because hybridisation yielding viable offspring can occur. Fertility can cross species level, for example, domestic dog with wolf or jackal, horse with zebra and donkey, radish with cabbage (both of the mustard family) and so on. Yet again, therefore, what is a species? And what a type or kind?

The genome of a eukaryote normally comprises 2 sets of chromosomes. If you double these (to 4 sets) you will produce no new genes but may have produced a 'tetraploid hybrid' that is fertile but cannot interbreed with with its parents' kind. This common outcome is classified as a new species. Moreover Gaston Bonnier, a botanist, planted the same lowland perennials in a lowland garden and in high mountains. He found that adaptation to the mountains (larger roots, tough hairy leaves, larger flowers etc.) converged on close resemblance to a previously-known mountain species. *There were, again, no new genes but speciation.* **As already noted, though, 'species' is an artificial distinction. It is a man-made not a natural division so that, in the face of such adaptive flexibility, it appears that the rank is less accurate a bio-sorting concept than 'type'[33] (which embraces species, genus and sometimes family within its limits).**

A corollary of this observation is that misconception has restricted the Latin meaning of 'species' to its current biological sense. In this sense the only possible origin of species is gradually due to random mutation and natural selection[34]. However, we now know that changing

---

[33] *SAS*: Chapter 22.
[34] see *SAS* Chapters 23: The Creator and 22: The Editor.

circumstances may precisely trigger different permutations of adaptive potential already codified in the in the genome. Non-random contingency plans are already embedded in an organism's program. *Thus the concept of naturally selected but preordained, codified response occurs.* **And, as a result, the concept of 'species' needs be re-expanded to its former, broader sphere of plasticity, the Latin type of species, 'type'.**

A second corollary is that any taxonomic system based on evolutionary relationships may, if macro-evolution is a false idea, not fully reflect the true nature of the living world.

Note also that, as with our own informative symbolism (called the English language), genetic language (written chemically on *DNA* as opposed to paper and ink) has grammatical rules which govern its employment. Furthermore, much of encyclopaedic *DNA* in the form of a genome is taken up with complex switching networks (in order that the right proteins are produced at the right times in the right places). This bio-chip (inside the nucleus of a cell which is in turn much smaller than a full-stop) is working incessantly and highly accurately in order to maintain health. Just as bugs in computer programs can cause a cascade of errors so can bugs in integrated genetic systems. Almost every cell of every organism is run by complex programs, code and nuclear bio-computing.

Within the compass of such a vastly complex and integrated instruction manual specific sections the store-and-print-efficient molecule called *DNA* (deoxyribonucleic acid) are transcribed into another sort of nucleic acid *RNA* (ribonucleic acid) which is then translated into the non-symbolic 'reality' of protein by a translator machine called a ribosome. Other machinery performs editorial checks to make sure the copy is accurate and errors minimised. Such protein constitutes most of your body including its sets of chemical master-agents called enzymes).

We all know wind, light and rain cannot produce such symbolic yet realistically dynamic text by chance but, as its mind-set is materialistic, the scientific community often deceives itself they can.[35] *Thus random mutation (accidental damage to DNA textual sequences) checked by 'death to the weakened' (natural selection) is held to be Generator Of Diversity.* **This alone can gradually create new species, genera etc. Thus it must have turned apes into men.** How reasonable is that? And yet to query this invokes insistent philosophical taboo.

Unveil pretence! Cast taboo into the fire of debate. *Could ape, even in the space of a couple of million years or about 10,000 generations, have plausibly 'slid' into man? Or is such gradual slip unnatural and disallowed?* **Can facts falsify one reading of the two?**

---

[35]   *SAS*: Chapters 20 and 21.

10. Man as Opposed to Ape

Before another plunge into the world of skewed interpretations, let us land from flights of fancy and inspect the neutral facts.

To watch a troupe of langurs hopping across the road from one belt of trees to another in the misty Himalayan foothills is to know that primates and man are akin. Closer still, the hominoids form a group of similar or homologous types. It is no surprise that, on the basis of thematic design, apes resemble man in most features of anatomy, physiology and biochemistry. In respect of some proteins for which sequence data has been run, the likeness between chimpanzee and man is quite close. But a man is a man and a chimp a chimp, for all that.

Essential non-molecular differences between man and ape involve:

1. Genetic program
2. Cranial capacity and morphology
3. Face, jaw and teeth
4. Arms and hands
5. Legs and feet
6. Pelvis
7. Language, literacy and numeracy
8. Metaphysical chasm

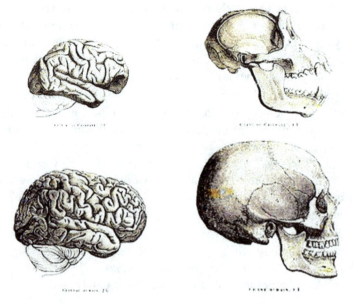

10.1   Paul Gervais: 1854 comparison of chimp and human brains and skulls (rendered to approximate scale).

46

The genetic program for all humans, regardless of race, is 99.9% the same. Different by a thousandth. The chimp (*Pan troglodytes*) routine is routinely retailed at 98.5% homologous; this amounts to about fifteen times further apart than those in the human ring are from each other. Put another way, this amounts to a difference in over 40 million base pairs that would need accurate, progressive alignment in mother sex cells by evolution in at least one hereditary line over the space of a mere 5 my or so to produce a human. Ain't Lady luck done well!

Certain aspects of life on earth (e.g. respiration, cell maintenance, reproductive routines and so on) are common; thus, on the simple basis of genetic sequence data, the banana, a plant, nears 60% human; and 75% of human genes have counterparts in a nematode worm which is not three-quarters human just because of that.

Although about 2% of the human genome codes for protein (which you might expect to be very similar in ape and human), about 98%, once considered 'junk', is known to code, along with the epigenome, for an extensive array of switches and controls.[36] This non-protein-coding *DNA* is where the real differences might be expected to occur.

Nevertheless, with respect to comparative analysis, various protein-coding fragments (about one three-thousandth of the whole chimp genome) were chosen because of their similarity to ours and assembled in an order based on the notion of evolutionary relationship. Roy Britten, the original estimator, has now lowered his equation to 94% but the similarity may well turn out at least as low as 90% if you count the non-protein-coding 'junk' and epigenetic factors as well as genes. Such non-coding elements are now recognised as critical regulatory and structural factors in the operation of *DNA*, ones which may well help to explain the differences between the two types of body. The mode of sequencing, where protein-coding sections comprise 2% of the total *DNA*, also excluded mention that over 50 human genes are partially or wholly absent in a chimp; it excludes both 'indels' (large sections of complete dissimilarity in non-protein-coding sections of *DNA*) which exist in the cells of humans but not the monkeys they 'came from' and *vice versa*; excludes mention that in evolutionary terms women are more like chimps than men (the Y chromosome has 'diverged' markedly more than the X); and fails to mention that a chimp genome is about 10% larger than human. In fact over 99% of mouse genes (about 30000 like us) have human counterparts. To summarise, a maximum 95% sameness equates to nearly 100 million specifically required *SNP*s (single nucleotide differences). Moreover, only if a mutation involves a sex call can it be inherited.

---

[36] see also Chapter 22 and *SAS* Chapter 23: Non-protein-coding *DNA*.

Within a postulate of only five million years or so such a chasm therefore rules out the gradual evolution of an ape to man.

The basic *brain volume* of normal humans varies from about 850-1450 cm³ except for pathological discrepancies (macrocephalic above 1500 cm³ and microcephalic less than 650 cm³). A child is around 370 cm³ at birth, 850 by four years and 95% of adult at ten. There exists wide variation in a normal population but the adult male norm is about 1350 cm³ and female about 1225 cm³ (or 10% less). Ape skulls vary from about 90-750 cm³. Gibbons are only about 100cm³. On average chimp skulls range between 250 and 500 cm³. So do orang-utan but gorilla skulls are between 340 and 750 cm³. The latter (and also an alleged hominin called *Paranthropus aethiopicus*) possess a bony, sagittal crest on their calvarium, which man does not. Given wide variations the possibility of an overlap between human and ape skulls is possible. It is easily possible, having reconstructed a skull from fragments, to produce a higher or a lower volume than the original whole might actually have enclosed.

In short, ape cranial capacity (including *Australopithecines*) ranges between about 275 to ~500 cm³ (largest chimp) and, for the largest gorilla, ~750 cm³. Meanwhile, across a large gap unless men evolved from gorilla-types, the *Homo* set ranges from about 800-1225 cm³ (*erecti*), 1200-1650 cm³ (Neanderthal) and 820-2200 cm³ (human). The implications of this gap and the true nature of such fossils as Java Man, Wadjak skulls (~1000-1500 cm³), *Homo rudolfensis* (~780 cm³) and Turkana Boy (reconstructed adolescent's skull ~900 cm³) are dealt with in following chapters.

You might, at this point, question whether skull enlarged before its brain (leaving vacant space) or *vice versa* (when skull would squeeze enlarging brain to death). Or, perhaps, with serendipity the factors co-evolved. That has to be materialism's mind-set's guess.

In fact, it is not so much brain size *per se* which is important. A small watch works as well as a large clock, a mouse is as clever as a rhinoceros and a woman no less intelligent than a man. What appears to count is the ratio of brain to body size. Where it suits, cranial capacity is taken as an indication of intelligence and skulls are arranged in ascending order of size. But it is another story in the case of Neanderthal man, whose cranial capacity was, on average, equal to or greater than that of modern man. Indeed, the large size of man's brain in 'savages' was one reason why Wallace, Darwin's partner in 1859, later distanced himself from Darwinism.

Apes all show 'facial prognathism' or protruding *jaws*, prominent brow-ridges and no temple (forehead) or chin. The jaw is U-shaped (human more V-shaped or 'parabolic') with teeth of the same number although of different shape and size. Man has small *teeth* with short eye-teeth (canines): apes have large teeth for masticating raw food and pronounced eye-teeth with which they fight. Man eats meat (but can well survive on a varied vegetarian diet); apes generally do not. Some Bonobos hunt but it is purely speculative why an ape should have modified a successful dietary habit to scavenge or hunt. Only theory requires it. An ape's lips are thinner and more mobile. Man's nose is more prominent and his nostrils less flaring.

Ape **arms** are long as befits athletic 'brachiation' i.e. pendulous, swinging motion through the trees.

As regards **hands**, how provident that fishy *Eusthenopteron* happened to have two bones instead of one in part of its fin. Did they evolve into the radius and ulna, which form an ingenious joint permitting rotation of the hand around its long axis? Or was this facility, basic to skilful manipulation, part of the plan for mammals and thus, optimized, for man?

Certainly, arms with hands are extensions that translate desire into practice. Hands, like 'outer brains', are manipulators of objects as opposed to symbolic thoughts. They reflect the desire of mind to rearrange the world according to its informative patterns. Man is conceptually creative and his very flexible hand supports this capacity. An ape's opposable thumb is set much closer to the wrist so that the ability to grip (except for branches) is reduced. However, man's flexible thumb enables him to manufacture as well as use complex tools; and the area of his cortex devoted to hand movements is relatively much larger than that found in apes. While the ape cannot move its fourth and fifth digits independently, man manipulates fire, machinery, electricity and, now, nuclear power. Evolution has not trumped creation with a suite of intermediate ape-to-man hands.

Legs and feet are, in humans, lower extensions that translate desire to relocate the body into (normally) bipedal practice. They represent the capacity for energetic locomotion. It may be significant that man has legs longer than his arms. Man's torso is shorter than the apes. His posture is upright with hands free while moving; although short-legged apes sometimes walk erect they mostly 'knuckle-walk'.

Man's specialized foot for walking upright is fully expressed even in the embryo. It is arched both across and lengthwise but cannot grasp objects. In fact, the foot of a mountain gorilla from Zaire differs from the hand-like foot of other apes, resembling that of a human. Its arms are not too long, nor its legs too short - a young gorilla can rear up and walk in a human way, resting on the sole of its foot rather than the side. But chimps and lowland gorillas have typical hand-like feet: such four-handedness includes toes like opposable thumbs and curled-down toes whose grip is well suited to swinging in the rain-forest. In this case no branch of factual evidence supports just-so proposals for the staged evolution of a flatter, walking-suited 'human' foot.

An ape's **pelvis** is not designed for the upright position; man's is flatter. His hip-bone is short and broad, the ape's long and narrow. This difference in shape is related to the curvature of the backbone which, in man, sweeps in an extended S to facilitate the upright posture. The human hip socket is to the front which also promotes a stable upright position while an ape's pivot is to the rear and so causes the creature to stoop. A half-way, evolutionary position would lack stability and impede both stooped and upright motion. As well as knee-joint, rib-cage and spine the *foramen magnum* (where spine enters skull) reflects these postures by its different position in each type of brain-case.

Why interesting? **Neoteny** is a process whereby juvenile traits of ancestors are, it is supposed, retained into the adult stages of development in the more 'advanced' kind. For example, newborn and juvenile chimps exhibit skulls considerably more rounded and 'human' in shape than adult chimps. In other words, the juvenile skulls more nearly resemble those of modern man than they do those of their own adult forms. It is possible to parade sequences of skulls which illustrate this fact.

As A. Montagu wrote: "In surveying the development of the concept of neoteny it becomes clear that the morphological changes in the varieties of humankind have been mainly brought about by the retention into adult life of traits principally characteristic of the foetus."[37]

Certainly, D'Arcy Thompson's theory of transformation shows that a regular mathematical relationship exists between certain forms - as if, for example, a shape on a rubber sheet were transformed by pulling the sheet in one direction or another. Whether this has anything to do with evolution is an entirely moot point.

Nevertheless, in a nutshell it is suggested that modern humans evolved from apes by retaining juvenile traits. It could contrarily be argued, although not by an evolutionist, that the similarity between adult human and juvenile ape skulls indicates that, in this case, the human, not the apish, type is basic. In other words, that recapitulation by acceleration occurred in the ape, so that pongids radiated from a human ancestor!

A Finnish lecturer in palaeontology has fuelled the debate. For him both contemporary and fossil forms disprove the standard notion that man is descended from the apes. It is well recognized that man is a generalized (though advanced) organism. Inhabiting a body capable of many activities, he is very adaptable and can make himself at home almost anywhere. Kurten points out that specialized dentition in apes (such as their premolars and 'fangs') makes it likely that their ancestor was a primitive creature with human dental traits. He suggests, drawing evidence from the Fayum fossil assemblage near Cairo, that the lines of man and ape separated nearly forty million years ago. It 'goes against

---

[37] Montagu A. *Growing Young,* McGraw Hill, 1981, p. 16.

the grain' that the small teeth of a cat-sized primate, *Propliopithecus,* would have evolved into the ape speciality and then back into man-like teeth again.[38] So Kurten suggests, after the end of the Oligocene period, two separate lines of descent, one to apes, the other to man. He also says that there have been a number of different species of man. Such species may, he conjectures, have coexisted but all except one, *Homo sapiens,* have now become extinct.

Why were *H. erectus* and Neanderthals illustrated as **hairy**? Because, of course, they were nearer apes! This assumption is as theory-driven and unproven as those involving intelligence.

Loss of fur causes exposure to cold, harmful U-V radiation and scratching by foliage and so on. Humans have as many follicles as apes but their hair shows, shorter and finer. Such disadvantageous devolution is, of course, offset by thermal controls and pigmentation (e.g sweating and dark-coloured skin); such factors may be less a product of evolution and more nuances of adaptive potential.[39] Hair shows slightly differently in men and women but its comparative lack on the not-so-naked ape enhances the presence of bare skin. However, although sexual signalling has been proposed as a reason for our 'nakedness', neither ape nor any other type animal seems to lack appropriate communication.

So socio-psychological excitement speculates verbosely how we came to hairy pate, face, armpits, genitals and seemingly 'naked' skin; top-down. Perhaps *atavism* proves that we evolved? For example, some completely human humans suffer from 'hypertrichosis' to the extent that their skin is covered with hair as thickly as a cat, dog, bear or chimpanzee. The condition is interpreted, theoretically, as a reversionary, single-character throw-back into long-lost ancestry.

Other well-known, *atavistic conditions* include the human 'tail', claws on the wings of some chick and adult birds and 'hen's teeth'. For example, an interesting experiment took outer tissue from the first and second gill arches of a five-day-old chick embryo; it was combined with inner embryonic tissue (mesenchyme) of a mouse taken in this case from the region where the first molar teeth develop. Normally the enamel layer of a tooth forms from outer tissue and underlying dentine and bone from the mesenchyme - *only if* that tissue can interact with the outer tissue. Chick mesenchyme cannot form dentine so that its outer tissue never gets the chance to form a tooth - but in the experiment where it was artificially exposed to the dentine-producing tissue of mice embryos, it did. And formed teeth! Teeth in a bird!

---

[38] Kurten B. *Not from the Apes,* Gollancz, 1972, pp. 38-40.

[39] *SAS* Chapter 23: Super-codes and Adaptive Potential.

In this case an unnatural construction protocol has liberated the potential for tooth formation. This suggests that an original modification to the genetic program for vertebrate mesenchyme has, in birds, disconnected it from the production of dentine and, therefore, teeth. If, as in the experiment, it can be reconnected it will produce teeth. In other words, birds have the genetic potential for teeth.

Let us elaborate. In a *top-down* view teeth are a subset, a subroutine associated with the archetype for vertebrate structure and switched on by chemical resonance (here by a chemical trigger present in embryonic mouse mesenchyme but absent in a normal chick). Such subroutine is relatively expressed or suppressed in the context of a particular organism in which the larger 'alimentary' routine from vertebrates is present. In other words, a principal or general vertebrate routine is refined according to its specifically intended vehicle (or body).

We know that in the past some extinct fish had jaws and armour plating (like stegosaurus or armadillo now). A variety of birds had teeth. Could high-level, 'master' switches appropriately lock such subroutines? Could inappropriate potential be more or less firmly 'blanked out'? Genetic switching systems at all levels are under intense scrutiny and, as yet, far from fully understood. ***Top-down* predicts that, once unravelled, these will decisively demonstrate vastly greater informative complexity and subsequent variation on potential than could possibly occur by chance**.

Thus, the aforementioned condition of 'hypertrichosis' might equally derive from intrinsic adaptive potential. In other words, it could represent abnormal switching in a 'hair' subroutine. In this case a 'malign' mutation to the switching system might cause suppressors to be turned off; or an abnormal copy number of hair-building genes be induced per cell; or some epigenetic mutation open a hair sluice. *In short, from a typological point of view atavism is not the accidental 'abruption' of a long deselected past but irregular access to part of an archetypal routine locked for that organism.* In the chick's case the trigger was the abnormal 'chime' of specific dental protein from a mouse. The one for over-hairiness is not yet known.

A man's ***larynx*** is able to produce a much broader range of vowels and consonants than any other organism. Thus he can enunciate enough sounds, complex or otherwise, to clearly speak his mind. In other words, he can employ ***syntactical language***, often of more complex construction in so-called primitive societies than in our own. The study of the earliest known language, Sanskrit, revolutionized the study of language and grammar, giving rise to the science of comparative philology. Beyond the spoken, the written word has allowed man to develop the information banks which symbolically underwrite his technical and cultural achievements.

Of course, animals communicate; birds sing, bees dance, gorillas gesture and dolphins give whistles, clicks, creaks, squawks and other sound signals. And of course they can think. Cats, rats, chimps, pigeons squirrels, crows etc. are able to recognize and respond to, if not always solve, contrived problems in mazes. What is unique to humans?

Animal signals are of limited dimensions and message, basically emotional and concerned with distress, food, location, sex etc. No clearly defined linear string of words, with grammatical structure, has ever been heard from an animal. Furthermore no animal can express, and then re-express in another way, an abstract idea. There is no great difference between the sounds of geographically separated members of the animal community. Wolves howl no matter where you hear them. With man language, among equally intelligent groups, varies greatly, so greatly that each group's language may be at first incomprehensible to the other. Moreover, words are symbols: they are a kind of code. Animals cannot manipulate symbols in the way linguistic or mathematical ability requires. In this way, a great chasm exists between the mind of man and ape.

It is true that, given a human teacher, animals may come to understand words and sign language as well as gestures and tones of voice. Apes are intelligent, sensitive creatures which respond to human affection. But this does not mean that, any more than parrots, they would by themselves have developed complex systems of symbol such as alphabets or mathematical or musical notations. It does not make them potential humans, any more than their humanoid form need have inexorably evolved into man. Only theory requires that.

The imitativeness of apes is remarkable. The husband-and-wife team of Dr. R.A. Gardner and Dr. B.T. Gardner began their famous experiments with Washoe, a chimpanzee, in June 1966, in association with the University of Nevada. They knew earlier experiments had shown that apes have relatively little interest in sound-play. Indeed, a previous experiment with Viki, a chimpanzee, had highlighted the quite serious psychological barriers against apes vocalizing. After six years she could use only four words and a few other sounds. There was no comprehension if previously learnt phrases were rearranged. Also, a chimp is unable to make refined use of its vocal chords. So the Gardners chose American sign language, as used by many deaf people in America.

After intensive care by humans in humanly contrived circumstances and isolated from the distraction of chimpanzee company, Washoe mastered sixty-seven signs (some calculate more) and 294 two-word permutations of these signs. Visually guided imitation brought rewards; but human language depends largely on auditory rather than visual factors. The human child has capacity to imitate sounds but a chimpanzee's capacity is greatly inferior. Its auditory-based recall, essential to the acquisition of language, is less than a parrot or crow's and neither creature has the cognitive processes of a human child. Apes merely learn, from the start, parrot-fashion. Children follow a pattern of language acquisition and grammatical regularization. By the age of two they will have passed the chimp and go on to learn an alphabet, spelling and writing. They may learn different languages and scripts but the pattern is the same. Washoe was a chasm away from human children as regards language potential.

Can an ape create a sentence? Studies have yielded no evidence of simian fluency with grammar. Such instances of presumed grammatical competence have been explained adequately by simple non-linguistic processes. The mean length of a child's sentence increases, so does its complexity. Not so with apes. Apes can, with non-simian tuition, pick up vocabularies of visual signals but there is no evidence that they can combine such symbols in order to create new meanings. They cannot construct syntactic sentences with different permutations of the same words. In contrast, by the age of five children with no special training can construct syntactical language. They can answer simple questions dealing with space and time, the measurement of heat, differences in size and weight, comparatives and superlatives, regular and irregular verb forms, pluralizations, active and passive moods, past and future tenses and much more. Elementary mathematical, imaginative and conceptual abilities are emerging.

It is psychologist Noam Chomsky's belief that all human languages share deep-seated properties of organization and structure in the brain.[40] These 'linguistic universals' are, he assumes, an innate mental endowment rather than the result of learning. If the capacity for language acquisition is innate in humans, may we presume the other forms of human quest (science and technology, philosophy and religion, literature etc.) are also exclusively human?

Dogs have a keener sense of smell, elephants keener hearing and eagles sharper sight than man. Organisms are equipped for different roles; they express different characters in an ecological drama. In the play of life on earth, mankind, granted a keener intellect than all the rest, has thereby inevitably been cast in the lead role. For, above all reflex senses and responses, it is innovative mind which rules. The basis of human endeavour and achievement is language, both verbal and mathematical; in short, it is the ability to manipulate symbols. The holistic view, like Chomsky's, is that use of these two kinds of language is programmed into human being. Literacy and numeracy are

---

[40] Chomsky N. *Psychology Today,* February 1965, pp. 432-3.

archetypally endowed facilities. There is absolutely no evidence (except in just-so, evolutionary imagination) that they somehow evolved.

Language is the tool by which man thinks, grasps abstracts and general principles apart from their particular expressions. What creature, other than man, can argue theoretically, logically and, sometimes, constructively? By this faculty there exists, as invisible as mind in brain, a distinct, *metaphysical chasm* between apes and humans. This neither *DNA* molecules can create nor genetic, physiological, anatomical analyses let alone palaeoanthropological speculation perceive. In physical principle, apes and humans are biologically very similar. There exist minor differences such as small canines, the ability to sweat, female breasts, obligatory bipedal instinct with appropriate construction and so on. But man is not an ape with a few added tricks.

The deep chasm lies in intellectual capacity and, deeper still, your ability to transcend (although necessarily also retain) the material or 'lower' mind of instinct, memory, sensation and physical reaction. *The latter are essentially passive, uncreative faculties.* **Transcendence, on the other hand, involves a purified form of the immaterial element, information.** With purifying concentration this element becomes active, expansive and creative.[41] Human profundity is able to discern truths beyond the superficial, sensible aspects of material operation. We are able (and it is the underlying thrust of education to enhance this capability) to work with abstracts, concepts and the perception of universal principles. We may realise global, cosmic and even, at root, the illumination of spiritual truths. Such whole transcendence (*illuminatio*) and its lesser reflection (which our forefathers called '*ratio*' as opposed to lower '*intellectus*') are the sole reasons for our self-styled tag of 'wise' or '*sapiens*'.

In short, a human being is capable of extensive communication, logic and mathematics. (S)he craves information, innovates artistically and creates technologically in high degree. The orderly work is often very complex in its integrative understanding or construction. **Man, designer, is the mover of the world.** Such qualitative potential removes man as far from apes as apes from insects, worms or jellyfish. *Physical similarity is not, although materialism's skew would order otherwise, the same as evolution.* Indeed, it may have nothing to do with evolution; and, certainly, quantitative genetics scarcely explains human qualities and powers at all. **In fact, the facts suggest that metaphysically-adept humans be placed in their own *Phylum cognoscens*.**

---

[41] see *SAS* Chapters 5 and 6 especially Top Teleology.

11. Fossils (Not Molecules)

There are two main methods used, with sets of assumptions, to gauge what happened in the biological past.

Fossils (petrified impressions from earth's international library) are examined to discover where difference might exist between modern and ancient organisms. They constitute 'gross', sensible evidence for theories of origins; and they are dated using 'radiometric clocks'.

On the other hand, molecular bio-sequences (of *DNA* and protein) are used, where possible, to discover 'subtle', insensible genomic differences between ancient and modern counterparts. In ideal conditions of preservation *DNA* has a half-life of about 520 years. This means that, theoretically, a strand would have disintegrated after, at maximum, about a million years but have become illegible far, far earlier. Actually, for fossils older than a few thousand years the method is impracticable.

In this case comparisons of sequence are made between the genomes of more and less similar organisms so that, using a 'molecular clock', the time taken for one sequence to hypothetically to turn (i.e. evolve) into another is calculated.

Both methods employ sets of assumptions which affect interpretation of results. The overarching one is whether the organisms involved are thought to evolved in Darwinian mode or have been derived from archetypes (see The Skew). The former is the naturalistic and therefore scientific preference. If, however, assumptions at any level change (for example, fresh, contradictory information is proven) then, as in a detective's forensic investigation, interpretations change accordingly.

We'll look at molecular comparisons later. For now let's examine fossils.[42] **Welcome, therefore, to a Hall of Smoke and Mirrors - palaeoanthropology!** Here 'evolutionary progress' runs apace. Surely we'll cement materialism's case? Thus press the ape-man button! Shelves bulge, keys chatter and, as noted, artistic licence runs incessantly amok. Febrile imaginations have, for six or seven generations, now eagerly clothed 'bones of contention' with an idiotic scowl and lumbering, part-upright frames. From ever-shifting dusts of excavation swirl as many ghosts and theories as there are experts in the

---

[42] see also *SAS* Chapter 22: Types of Fossil.

field. It is entirely reasonable to claim that evidence remains disconnected, hard to decipher, often media-hyped and always hotly debated. Speculative tales proliferate; hypotheses and spats spill forth. Does academic politics or science win the game? Various arguable reconstructions of bone fragments have secured many a grant, degree and reputation - even fame and fortune but, still, does confusion or consensus reign? Institutions are established and their libraries of manuscript and file retro-cast the intellectual weather that surrounds proposed development of ape to man. And, of course, that mutant ape is you. You're a mutant, brother, and you, sister, too. We're all mutants, are we not? No problem, though, because a body is an animal and what are we but bodies working like computerised machines? I'm not worried I'm a mutant ape, amphibian or fish; and if my ancestry is thus why should I care if, in the last analysis, my line springs from a mutant micro-organism and at root I was a germ? How else were you supposed to come of age, to have developed into evolutionary maturity?

Science is a challenge. How can progress finish? But presentation's always positive; only when a new discovery is made, technology for measurement invented or fresh methodology developed is previous, underlying weakness superseded with a fresh wave of authority. Such flexible authority is, however, in its exclusivity inflexible. Its mind-set cannot flex beyond materialism's box. When it comes to monkey business an unyielding, 'scientific' skew insists that data be shoe-horned into the ape-to-man scenario. It assumes what it is trying to prove is true. *By such tautology fossils are, of course, the mark of evolution; but, remember, the abductive logic of a history book only yields best guesses.* The literature of palaeaonthropology is far too large and convoluted to engage with other than broad brushstrokes yet a balanced outline of both *bottom-up* <u>and</u> *top-down* perspectives must be drawn.

No doubt, *bottom-up*, evolution theory *must* turn apes to men. Indeed, a 'cast-iron' theory and dependent world-view hang upon this transformation. Substantiation is, therefore, imperative. Much is at stake. The whole materialistic world-view stands or falls. Therefore we seek and find. The maxim, 'If I hadn't believed it I wouldn't have seen it that way' seems to frequently apply. Maximum suggestion is squeezed out of minimal evidence. Fragmentary bones of contention are named, classified and re-classified interminably according to each 'revelatory' find. Compelling public statements are, one might reiterate, issued at the time of each discovery only, amid behind-the-scenes controversy, to be quietly dropped; retrospective ignorance (after further clues are found) is common and the whole subject has been implicated in degrees of skulduggery that range from outright fraud through secrecy, naïve wishful thinking and imaginary propaganda through to skulduggery's lack - an honest search to find, if they exist as surely they must, the relevant fossils.

Such a search involves the comparative anatomy of fragmentary, whole and reconstructed bones. It includes the reassembly of putative and mostly partial skeletons, supporting dates and, most problematic, forever one-track interpretations of the data. To question either data-crunching mind-set or its dating assumptions is, sadly, to court a real possibility of curt, contemptuous dismissal. However, while science engages an excellent and supposedly self-correcting methodology to deal with the current operations of all things physical (including biological), it relies upon abductive argument - best explanation by inference[43] - for historical interpretations of past facts, especially in the case of origins. And while the first and foremost requirement in an honest court of historical appeals should be for *both* sides to put their case, this is not the evolutionary determination. **To repeat, therefore, until the message resonates - fossil evidence should be subjected to both *bottom-up* _and_ *top-down* consideration. Holistic design and materialistic evolution should be thoroughly, consistently compared. No less than a faculty of Comparative Palaeoanthropology demands establishment!** Each side might learn something from the other. Instead high-pitched objections from materialism's camp pierce reason's lucid air. Only in a complementary way, however, can the smoke be cleared, mirrors shattered and fun-of-a-scientific-fair be elevated into wholesome dialogue.

Darwinian theory *needs* differentiation and division. To cement its graduated splits a Linnaean scheme of prejudice has been established.[44] Within it fossils are arranged and named then, often, re-arranged and re-named according to malleable permutations of the ape-to-man conceit. 'Hominin' is a designation invented to mean humans and their evolutionary ancestors. There is indeed no lack of fossil evidence for such creatures. The Natural History Museum in London published a Catalogue of Fossil Hominids. Although items rather than individuals are tabulated, a count of some 4000 individuals has been estimated (by Marvin Lubenow) to have been discovered up to 1976. This figure includes over 200 Neanderthals and over 100 *Homo erectus* designations. Since then many more fossils have been dug up so that, mostly in highly fragmentary form, perhaps 9000 hominids altogether have been found. The finds have not, however, been systematically committed to a single list. Moreover fragile, irreplaceable specimens

---

[43] see Glossary: logic.

[44] see Hominoid Classification Reference and Glossary: *Hominoideae* etc.

are often and understandably sequestered in vaults and strong-rooms (as in Kenya, Ethiopia and South Africa) and elsewhere (such as Indonesia) also barred from access except by a chosen inner circle. Some 'finders-keepers' will not allow even well-qualified co-workers in the field near their 'property' or allow publication of any evaluation that may contradict their own (and, it may be added, their own fame and fortune). This caper has, in the case of determined but defensive missing-link hunters, gone on since the time of Eugene Dubois in Java (1891). Since originals are practically unavailable for study plaster casts are, in a small percentage of cases, issued to replace them. In some cases these casts are accurate and in others less so. For example, a plaster cast of the fraudulent Piltdown Man issued by the National History Museum for inspection by an investigator (Louis Leakey) had teeth of the orang-utan mandible which showed no evidence of file marks crucially evident on the original forgery. In 1984 the American Museum of Natural History in New York sponsored, in part due to 'concern about creation science', an Ancestors exhibit. Although nations such as China, Kenya, Tanzania, Ethiopia and Australia demurred a 'family gathering' of about forty original specimens was displayed. These national treasures were placed on mounts that had been precisely prepared using plaster casts - although in most cases the real bones did not fit and the mounts had to be adjusted accordingly. How dependable is such a cast of players? To what extent is the whole business theatre?

The upshot is that, although many fossils exist, they are neither catalogued in a comprehensive way, adequately displayed or available for close study. Only a very tiny fraction of palaeoanthropologists, archaeologists, anatomists, evolutionary biologists, sociologists or any other interested party can gain sufficient access to make an independent judgment unclouded by expert opinion, academic fashion, second-hand media hype or other kinds of 'Chinese whisper'. Since when was hearsay how you did your science?

12.   Ages of Rock

Let's start with fossil dates and morphology.

For a number of convergent, abductive reasons current science believes that the cosmos is nearly 14 billion years old. One method used to make this deduction is extreme extrapolation from present circumstance. However everyone, not least a man of science, understands the danger of such extrapolation into future or past - because no way exists to test it; and that, if assumptions involved are disproved, conclusions will change. Thus such figure may or, in the future subject to revision, may not turn out true. The fact is none of us *know* for sure the age of cosmos or of earth.

However, let us grant we've got it right. Such age would supply an easy-going, happy-go-lucky theory of evolution with a plausible space of time. *On the other hand, whether earth is young or old, holism takes the line that information, not time and chance, is the critical factor.* **Informative mind precedes informed energy.** ***Cosmo-logical design trumps days, months, weeks or billions of years. And bio-logical codes drive the point right home. Life's times do not depend on mindless physic.***

Age alone cannot create a set of cutlery let alone the specified complexity of bio-forms. Physics shows initial projection so fine-tuned that, although including elements of chance (the fodder of a naturalistic view), design is a reasonable choice for cosmic origin. And the science of biology now shows (see *SAS* Books 2 and 3 *passim*) that naturalistic process is an insufficient 'author' of our human hours.

Thus, to reiterate, the main issue of creation is not whatever kind of time you choose but *mind*. Mind is natural. Creative mind is anti-chance and has its range of purposes; expressing these it gives or takes whatever time it likes. But one thing every creator strives to eliminate (with an effect proportional to his intelligence) is bugs, error and the chaotic incidence of chance-created damage. ***Only mind is capable of systematic determination, coherent construction and purposive complexity; the time it takes to realise its ideas is of secondary relevance. Of course, this bears directly on the study of highly informed, codified life-forms whose past fossils reflect.***

However, the fossil record is cast in, we assuredly believe, ancient rocks. For all its shortcomings, doesn't it represent to reasonable men powerful evidence that evolution must have occurred? So let's ignore the immaterial element of mind and at least pretend that a high-grade, high-volume wealth of information can be input randomly. Thus mindless evolution *must have* made those fossils, each and every one.

Modern geology is based on two major premises - that the earth has great age and that Hutton's principle of uniformitarianism (summarized in the aphorism 'the present is the key to the past') holds good.

And theory demands that every fossil[45] in an evolutionary sequence must have both the proper morphology (shape) to fit that sequence and an appropriate date to justify its position there. Clearly, since the morphology of a fossil cannot be changed, dating is the more subjective element of the two factors. Yet, accurate dating of fossils is so essential that the scientific respectability of evolution relies on fossils having appropriate dates. Often, however, the date assigned to a fossil specimen does not agree with the theoretical order in which its 'evolving' morphology 'ought' to be found. A typical response involves the reassignment of any age-awkward specimen to a more accommodating taxon (group of organisms judged to form a unit). A second common response involves reassignment of the date. How can this be?

It's not at all that anyone is trying to *deceive* but simply that the theory's fact and therefore facts must fit it. Professionals are paid to peer through evolutionary spectacles. No spectacles? You're fired! The pressure's therefore overwhelming. Anomalies must be resolved this way or that - but only in the framework that's prescribed.

Geology's (and therefore palaeanthropology's) stratigraphic column[46] is built up of rock units, for example, coal and chalk measures, and subdivided into beds and bedding planes. By using certain fossils as indicators, the column can be divided into time zones. This 'bio-stratigraphic column' is defined by zone or index fossils, which occur in certain strata. For instance, any rock containing fossils of one type of trilobite *(Paradoxides)* is called a Cambrian rock: of another type *(Bathyurus)* Ordovician. In this way a relative geological timetable is compiled.

Absolute dates, for example, 430-500 million years ago for the Ordovician period, are ascribed by using igneous rocks for calibration points. These may occur among the sediments in the form of sills and dykes, and radiometric methods can be used to date them accurately. Sediments on which the dated igneous rocks lie must predate them:

---

[45]   *SAS* Chapter 22: Types of Fossil.
[46]   see Uniformitarian (Evolutionary) Chronology.

equally, those overlying igneous rocks must be younger. So the geologist ascribes ages to all his strata.

No-one claims his understanding is flawless. While evolution requires long periods of time (and is the habitual mode of geological thinking), the creation of a series of machines may occur 'all at once' or be extended successively over a period of time. Many creationists, while conceding that the question of the age of the earth and the universe is an important one, believe that it is independent of the question of whether or not creation occurred. They insist, for reasons explained above, that time is not the critical factor and are prepared to suspend judgment concerning whatever shifting time-scales are fashionable. These have changed, doubling on average every fifteen years, from about four million years in Lord Kelvin's day to 4500 million now.

Both poet and atomic physicist will agree that metaphors are not to be taken literally. Creation myths, whether Biblical (from Mesopotamia), Polynesian, Scandinavian or whatever, may express the truth of creation in a simple, allegorical way - nevertheless Biblical literalists challenge orthodox geology in three ways. Firstly, it was noticed that geology relies heavily on the evolutionary interpretation of the fossil record because the relative ages of various rock formations are determined largely in accordance with the presumed stage-of-evolution of the fossils in them. 'The only chronometric scale applicable in geologic history for the stratigraphic classification of sedimentary rocks and for dating geological events is furnished by the fossils. Where there appears to be a discordance between physical and fossil evidence as to the age of any series of beds, fossil evidence is usually preferred - even to radiometric dating. There is no one area where the whole series of fossils is found; but an evolutionist, having decided that living things evolved from less to more complex, arranges finds from different areas on the assumption that his decision is correct.

Herein lies a powerful tautology, a circular argument. **The assumption of evolution is the basis upon which index fossils are used to date the rocks; and these same fossils are supposed to provide the main evidence for evolution. The fossil record, itself based on the assumption of evolution, is interpreted to teach evolution.** By this sort of reckoning, the main evidence for evolution is the assumption of evolution.

But several index fossils have, after an absence of millions of years, turned up alive. Examples are the tuatara lizard, the small mollusc *Neopilina galatea,* the maidenhair tree and the dawn redwood; most celebrated is the 'link fossil' coelacanth fished up off the coast of Madagascar in 1938 after seventy million years' absence. Any rock dated according to coelacanth index is not now seventy million years old but anywhere between a hundred years and seventy million. Because most index fossils are small marine organisms, for example, molluscs, and the ocean depths are mostly unexplored, it is possible more index fossils will be designated.

But even if not, it is of little consequence to the argument. Persistence is a challenge (for example, mosquitos and flies in amber dated 60 million years are exactly the same as mosquitos and flies today). Evolution has to explain not only persistent species but, much more common, persistence of families, orders, classes etc. As regards classification, all kingdoms and subkingdoms are represented from the Cambrian onward. All classes of the animal kingdom are represented from the Cambrian onward except insects (Devonian onward) and moss-corals from the Ordovician. All phyla in the plant kingdom are represented from the Triassic onward except bacteria, algae, fungi (pre-Cambrian onward); mosses and horsetails (Silurian onward); diatoms (Jurassic onward) and flowering plants (Cretaceous onward). As we noted earlier, such groups appear suddenly in the fossil record, with no indication of transitional forms from earlier types. This is the case even for most genera and species. Index-fossil sequences such as micraster, ammonites and trilobites indicate only variation ('micro-evolution').

'Christian fundamentalists' interpret geology in terms of a young earth. A global catastrophe, resulting in a flood, was followed by several thousand years of uniformitarian processes. For non-gradualists (catastrophists) the geologic systems represent broad ecologic and sedimentary zones. They include fossils of creatures that lived at different places at the same time. The geologic column therefore represents the usual sequence in which rising, turbulent flood waters buried plants and animals from these different zones. Deposition and flood currents would not be entirely predictable but normally soft-bodied bacteria and worms that inhabit poorly oxygenated depths below the sea floor would be discovered in the lowest strata. Shelly forms from the sea floor would be found next, followed by marine organisms such as fish. Later would come land-plants and creatures. Those best able to flee the waters would be last and least likely to be buried by rapid deposition of sediment and fossilized; and these would include highly mobile, relatively intelligent hominoids. Such prescription fits the fossil facts.

However, no less than the evolutionist model theirs has problems. For example, why do we find fossil successions? Oyster-like creatures are found from bottom to top of the record - strange for slow-moving bottom-dwellers. This kind of succession is particularly marked in chalk deposits where a definite succession of different species of the same type of creature are found separate and unmixed at different levels. Chalk cliffs are not easily explained by young-earthers. Nor are the layers of salt hundreds of feet thick thought to have been laid down as seas slowly evaporated. Nor has the case for a young earth using haloes perhaps caused by the very rapid radioactive decay of polonium in zircon crystals remained unchallenged. Nevertheless, 'young-earthers' keenly interrogate the assumptions behind radiometric dating. What, they ask, is the absolute validity of such a dating method?

'Absolute' radiometric dating (including isochron and Concordia methods) involves both strength and weakness of assumptions. No doubt physicists observe unstable atoms changing into stable at certain rates and then extrapolate into the unobserved past. How, though, do you calculate how much stable isotope or how many molecules of the same type as decay products were already present at the time a rock solidified? How can you know that the decay rate has always been the same or that no contamination has occurred?

The age of a rock is defined as the moment it solidifies. The problem is to find out when that was. This is exemplified in recent cases such as the construction of Surtsey Island, the destruction of Santorini and the eruption of Mount St. Helens in Washington State (1980). But for now consider the actively volcanic Mount Ngauruhoe in New Zealand. Lava samples from a series of eruptions between 1949 and 1975 were tested 'incognito' by Geochron Laboratories, Cambridge, Massachusetts. Four of these were dated, with a 20% margin of error, at less than 27 kya (27000 years ago); at the other end of the scale one was assigned 3.5 myr (3.5 million years)! **This error represents a span practically as large as the whole of alleged human evolution!** No doubt, one is informed that radiometry does not work well with rocks less than fifty, or even a thousand, years - which human witness accounts might calibrate; but where it *is* supposed to work it can't be double-checked. A best attempt is made but no independent verification is possible. Thus great ages constitute an unchallenged and, it might appear, unchallengeable assumption.

If the potassium/ argon method can transmute a known 50 years into a wildly inaccurate 3.5 million then rubidium/ strontium yielded an age of 133 million years, samarium/ neodymium 197 million and uranium/ lead 3.9 billion years! And Grand Canyon volcanic basalt flows over the top of the North Rim measured by potassium/ argon have yielded (at one million years) dates more than a thousand times less than by rubidium/ strontium (1.14 billion). This latter is the same age as basalts buried deep under the eastern walls. Top and bottom the same. But each 2.6 billion less than their age as indicated by uranium/ lead decay! It is as if one system aged you 20 and the other 45!

As regards the alleged evolution of some kind of ape into human form, the prevalent use of potassium/ argon and carbon-14 dating, including the '**coverage gap**', need special mention. The effective range for K-Ar (potassium/ argon) dating is thought to kick in at about 200 (or some think up to 400) kya. On the other hand, the useful range for carbon-14 (whose half-life for decay to nitrogen-14 is 5730 years) is limited to about six half-lives, that is, about 40 kyr. *Such a gap (~400 – 40 kyr) means that during a critical, up-to-360-kyr period of human evolution involving many hundreds of fossils dating has been provisional, uncertain and even, for those chaps dug up predating radiometry, absent.* This is not conducive to accurate chronology but a method for precisely measuring (to 0.001%) the ratio of carbon-14 to carbon-12 in a sample against that found in the modern atmosphere has been developed. It is called an accelerator mass spectrometer and theoretically extends the range of carbon-14 dating to about 90 kyr (still leaving a gap of ~310 kyr). However, the real surprise is that no fossil material anywhere can be found with as little as 0.001%; and the consequent implication is that, because all carbon-14 should have disappeared by maximum 250 kyr, no fossil is older than that. *Thus great age would appear to represent, without the invention of significant excuse, an unutterable fiction.* Some mistake, surely?

How does palaeoanthropology respond? **When theory's right the facts must not be wrong!** Thus it reacts either by re-assigning morphological taxon or revising the date. Often the date assigned to a fossil specimen does not agree with the theoretical order in which its 'evolving' morphology 'ought' to be found. A typical response, due for example to the sparse and fragmentary nature of fossil bones, is to re-interpret geological data or 'evolve' a new evolutionary model. It may also, as previously noted, involve reassignment from the 'wrong' taxon to more satisfactory one. Such reassignment, which can ignite acrimonious academic spats, occurs quite frequently in palaeoanthropology.

A second kind of response involves reassignment of the date. How can this be? **It's not at all that anyone is trying to *deceive* but simply that, as always, theory's fact and therefore facts must fit it.** Whether or not this amounts to self-deception, professionals are paid to peer through evolutionary spectacles. No such lens? You're fired! The pressure's therefore overwhelming. Anomalies must be resolved this way or that - but only in the framework that's prescribed. Thus misidentification, 'intrusive burial', hoax or other reason is inevitably prescribed to dismiss a human fossil found 'too deep'. For example, the Calaveras skull and Castenodolo skeleton receive such proscription.

The 'coverage gap' can also provide date-flexibility. Thus, as regards the highly charged arena of hominid debate, large imprecision has led to a catalogue of reassignments according to where and in which scheme of evolutionary events a fossil of uncertain date is deemed 'rightly placed'; and also, through such manipulation, what deep ancestral relationship different academics might, in terms of family 'bush' or 'tree', assign to us. In every instance, when theory's right the facts must not be wrong! *And yet, whatever the age of our spaceship, it may be shown that precise, theory-concordant dating for many hominid fossils is lacking.*

Radiometric methods aren't, we've seen, rock solid. But this is not the point to enter a debate over chronology and what must, for evolution, be a very old earth. *It is, though, fair to point out that faulty assumptions are bound to generate faulty results. And to note that palaeoanthropology is not, it may be plausibly substantiated and thus fairly admitted, the most exact or easily self-correcting of sciences.* **Finally, when all is said and done, is it even possible that enough evidence has already been collected to falsify, whatever other fossils are later found, the ape-man missing link hypothesis?**

In this case can the diagram below be true? **Could it be possible that fossil evidence shows the ape-man saga as a myth?**

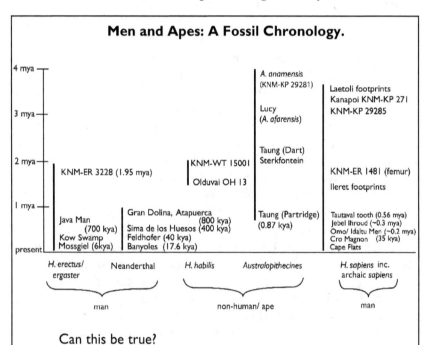

Can this be true?

This chronological chart represents a précis of fossil facts drawn from the thousands of 'hominid' fragments now excavated. Only a few, including the oldest and youngest that bracket each group, are shown. While the chart reflects fact for bones and consensus estimate for age it does not necessarily reflect such multiple and often heatedly conflicting interpretations, manipulations and hypothetical organisations as impact presentation of the actual material. Finally, it excludes from consideration controversial items such as the Piltdown skull, *Hesperopithecus*, the Castenedolo skeleton and so on.

As far as Natural Dialectic is concerned the chronology emphasizes a pressing need to drop philosophical prejudice and enjoin comparative anatomy with respect to the measurement. By this is meant the engagement of expert palaeontologists of both 'scientific' and 'unscientific' mind-sets (that is, arguing from evolutionary 'design-by-non-design' and non-evolutionary design respectively); and, in this dual/ duel engagement, rigorous examination and interpretation of both 'hominid' and all other fossil material. Only with such discipline, as in a court of law, will truth obtain a fair trial. Who shall be the jury but your public self?

12.1

A theory's scientific when it can be falsified. Can evolution's tale be falsified by fossils? If it can, this diagram demonstrates it is. And the story that's been built up since the time of Darwin isn't true. But, if you believe that evolution's true, it can't be right. It's wrong. The fossils and/ or dates would need a rapid reconstruction.

Let us see. Combine dates with morphology. Examples of fossilised *H. sapiens* range from about 5 kya (Cape Flats, South Africa), 25 kya (Zhoukoudian, China) and 30 kya (Cro-Magnon, France) through 60 kya (Lake Mungo, Australia), 100 kya (Skhul and Qafzeh caves, Israel), ~110 kya (Fuyan cave, Daoxian, China), 700 kya (Java Man, human femur), 1.9 kya (KNM-ER 1472, 1481, femurs and other fragments, and possibly 1470, cranium, Kenya) to 3.7 mya (KNM-KP 29285 tibia ends and KNM-KP 271 upper arm, Kanapoi, Africa. To this range might be added the modern-looking footprints from Ileret in Kenya (1.75 myr and thus dubbed *H. erectus*) and Laetoli tuff., Tanzania (3.75 myr and so deemed *Australopithecus afarensis*).

No doubt, to believe you are a modified ape requires you to accept thousands, if not millions, of unlikely propositions even as to what occurred before the ape. After this, how was the ape-man scenario established such that the world seems now enchanted by mythology of evolution and its history of man?

The rest of the book will treat each of the five vertical bars separately. It is dedicated to the proposition that they better

represent the fossil and molecular truth than myriad theory-driven speculations, classifications and family trees that, in different degrees, misrepresent it. Indeed, such misrepresentation has, on occasion, been deliberate so that the theory of evolution might be internationally proclaimed. More often the pseudoscientific manipulation of data has been unconscious because, with skewed mind-set, researchers are convinced their perspective is the right and only one. The range of presentations therefore, as we'll see, run a gamut between fraud and self-deception. Let us plot the course of mythic distortion. Let us break the spell.

13. Years of Palaeoanthropology

A brief history of palaeoanthropology both sets man's prehistoric scene and illustrates a series of events in the momentum-generating development of the theory of evolution with, importantly, a concomitant rise of humanism, materialism, atheism and, in the 'modern' west, the wane of formal religion.

Palaeoanthropology is, like all professions, guarded by its 'legalese', that is, by sophisticated bluster serving both academic precision and the qualification of 'insiders'. A determined student, having overcome this challenge of vocabulary, then faces a plethora of fragmentary material with which, according strictly to an evolutionary framework, a blur of multiple and often conflicting interpretations, suggestions, inferences and hypotheses is worked. Who, when bony scraps kick up such hot-aired clouds of dust, can tell the mirage from reality; who can see the wood for trees?

In this respect and at this point it is worth remembering skew, imperative and the fact that materialism may, in its correctness concerning physicality, still err with respect to the whole truth. A metaphysical element of information may play a critical role in the nature of cosmos and the source of coded life.

We respect fair and honest scholarship. Do we always find it in the Ape-Man Saga which, remember, might be hooting up the wrong ancestral tree? If not, how were truths of science established in the public mind? If so, how was the momentum of mythology built up?

It began in 1859 because, until then, you weren't even potentially a mutant ape. Charles Darwin was not aware of any ape-man fossil but, after his Origin of Species it was obvious, if organisms evolved, that humans evolved from 'lower forms'. It seemed, from similarity of form, this must be apes. Thus, for the first time in history, you theoretically became a mutant and, off the blocks, the race was on to find what must (if evolution is a fact) exist - evidence of your ancestral 'missing links'.

"On which side - grandfather's or grandmother's - were ape ancestors to be found?" asked Bishop Wilberforce in Oxford. "On every side", answered the Darwinists "we'll comb the fossil record and find them." What were they looking for? The break between modern man and modern apes is complete, but at some time in the past must there not have been ancestors that were half ape and half man? Distinguished amateurs deceived professionals. Distinguished professionals deceived themselves and even a Jesuit priest was caught up in the monkey business surrounding the search.

The international search for what was considered by the theory fact, ape-men, will be outlined below before each strand is dealt with in more detail.

The Feldhofer skull,[47] found in 1856, was at the time judged human. Thomas Huxley, Darwin's 'bulldog', knew about but could not plug it into the ape-to-man sequence for which fossil evidence had already become gold dust. Do you remember Howell's fake parade? This, the frontispiece of Huxley's Evidence of Man's Place in Nature (1863), is where the inspiration's from. However, at the time this Neanderthal skull was considered a human remain from the Noachian flood.

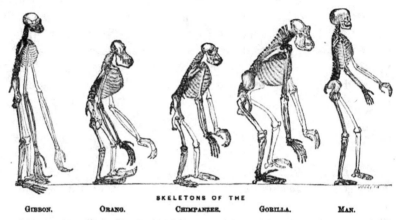

SKELETONS OF THE

GIBBON.   ORANG.   CHIMPANZEE.   GORILLA.   MAN.

Photographically reduced from Diagrams of the natural size (except that of the Gibbon, which was twice as large as nature), drawn by Mr. Waterhouse Hawkins from specimens in the Museum of the Royal College of Surgeons.

13.1   Huxley's Ape and Man Parade

In 1890 the *Pithecanthropus/ H. erectus* saga began with Java Man.

In 1907 an 'archaic man', in the form of a jawbone, was found at Maur in Germany and called *Homo heidelbergensis*.

Not to be outdone, in 1908 British pride was re-inflated by a find classified as *Eoanthropus dawsonii* aka the Piltdown fraud.

Also in this year Marcellin Boule from the National Museum of Natural History in Paris described the almost complete skeleton of a Neanderthal ritualistically buried near La Chapelle-aux-Saints in the Dordogne, France. Ignoring evidence of severe deformation by arthritis and rickets Boule interpreted the bones as more ape than human, not least in an ability to walk erect. Although incorrect, Boule's prejudice informed the public's mind-set concerning *Homo neanderthalensis* for fifty years or more. It generated countless 'artistic fleshing out' whose effect lingers to this day.

---

[47]   see Neanderthal.

From 1912 to the 1940's the *Pithecanthropus* saga continued with Peking Man.

In 1921 the skull of rapidly-aging Kabwe, Rhodesian or Broken Hill Man was hauled from a lead and zinc mine and a report on its 'savage' morphology authored by Arthur Smith Woodward of the British Museum (Natural History).

In 1922 a tooth was found in Nebraska and with premature zeal classified as *Hesperopithecus haroldcookii* (Harold Cook's Evening Ape). Later investigation proved (see *Cartoons*) that the tooth was that of an extinct pig. Scientific self-correction occurred but not before the media horse had bolted. Primary misinformation had hyped evolution theory across the world and had time to successfully service (in 1925) the Scopes Trial.

In 1924 'Dart's baby', the Ta-ung (or Taung) skull, was discovered in South Africa. The notion of a half-breed called *Australopithecines* was conceived. This important transitional notion has continued and continues in Southern, Eastern and North-eastern areas of that continent.

Australopithecines have nothing to do with Australia but, since the 1960's a stream of interesting fossils (including the Mossgiel cranium and Cossack skull) have been unearthed there.

In 1964 Nature announces *Homo habilis*.

In 1965 Howell's 'March of Progress' was published and since exploited by the full range of informative technology. Who says evolution isn't true?

In 1974 Lucy came to light.

Since this time a continual stream of bone fragments have hit the headlines 'proving' one palaeontological model or another. A controversial one was the hobbit, Flo or *H. floresiensis* (2004). Palaeanthropology often humanizes fossils, human and ape, using theatrical names. Another, more recent, is *Homo naledi*.

By the 1970's molecular biology was growing fast. A possible way to date fossils was discovered using mitochondrial *DNA*. Indeed, in 1997 such *DNA* was extracted from the right arm bone of the 1856 Feldhofer discovery!

Before this however, by 1987 strong focus on African excavations had veered to the pronouncement of an African Eve, Out of Africa or Mitochondrial Eve model which has, to this day, dominated palaeontological theory and thus media repetition.

Lack of imagination may allow us to take the discoveries that have transformed technology for granted. It is, however, these alone that make life different from how our forebears had to live.

Therefore, before closer study of each abovementioned anthropological characters, try to imagine what is perhaps impossible. Imagine that, with modern education and all knowledge of technology erased from your mind, you were transported to a large, wild location devoid, except for a few families, of other human habitation. **How would *you* survive?** How do 'stone age' people today survive? They live and travel light. Do not expect, after a year or two, to find much trace of their abodes. And if, in northern climes, a cave gave shelter in fierce weather would you be surprised? Those versed in jungle, desert and other survival techniques will tell you that such band of humans would certainly be stretched. How would you behave? Ask yourself whether any tools that such a band might leave do not correspond to what you might expect to invent as sensible, skilful starters in such arduous training as you yourself would, in that pristine situation, have to undergo. And is such kit still used by humans identified by science and society over the last couple of centuries as 'primitively savage'? Yet aren't bushmen, pygmies and jungle dwellers fully human? Shouldn't your own learning process have labelled you, in its technological infancy, half ape?

A final point. How should a book that criticises, on archetypal principle, the Darwinian version of man's origins be organised? Should the characters that populate its narrative be arranged according to the palaeoanthropological discipline's own history - as just described?

Or, since such description involves jumping up and down the supposed evolutionary progression, should the narrative follow a straight earliest-to-latest time-line? It would then describe the characters that populated blocks of evolutionary time but, because each one (e.g. Neanderthal or *Homo erectus*) might span several blocks the method would be repetitious.

In this case wouldn't it be best to take each character, from earliest to most recent, and deal with it throughout its span starting with, say, *Graecopithecus freybergi* or *Sahelanthropus* and working through to Cro-Magnon and modern man? Except that, for example, Cro-Magnon does not slot neatly between Neanderthal and *H. sapiens*. And except that (see Chart in the Chapter 'Ages of Rock: Man and Apes, a Fossil Chronology) dating is often far from exact. Nor is the progression smooth. The timeline of one type does not start where the other left off, proposed spans of existence for the characters vary so that overlaps of differing length occur.

It is also, fortunately or unfortunately, necessary to take into account the non-evolutionary notion that no such progression, only variation on an archetypal theme, ever really occurred - which throws a spanner in the works. Confusion, apoplexy, differing assessments and resulting complications then play their part.

So I have chosen to follow the course of palaeoanthropological history until the second world war and deal with each character in the order of its scientific discovery. After the war things become, with more workers in the field, more complex. Finds worldwide are covered although, through the Leakey family's work, focus is now concentrated on Africa as 'the cradle of mankind'. Various theories seeking to locate human emergence are described and criticised. These include an account of the recent impact of genetics before, at last, we both turn to Australia and return to Europe for the summing up.

14. Neanderthal

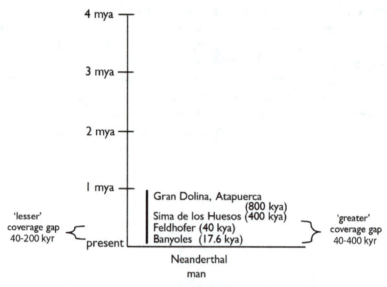

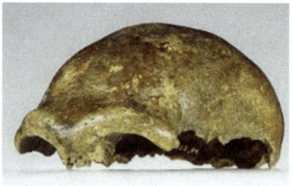

14.2   Neanderthal 1 (1856) skullcap.

It is usual for a crucifix, raised in a conspicuous position, to beam its message to the villagers of France. But beneath an overhanging cliff, above the inhabitants of Les-Eyzies-de-Tayac in the Dordogne, there looms another sign. No matter that cave paintings in next-door Lascaux involve artistic capability, across from the Les-Eyzies National Museum of Prehistory the statue of a brute Neanderthal silently 'preaches' ape-to-man evolution.

Joachim Neander (1650-80), foremost hymn-writer of the German Reformed Church, unwittingly gave his name to Neanderthal man. Neanderthal is a limestone gorge, near the village of Hochdal between

Dusseldorf and Elberfeld, through which flows the Dussel river. Neander's love of nature used to lead him to this ravine where, it is said, he composed many of his hymns. Almost two centuries later workmen, quarrying in the valley, uncovered the skullcap and several other bones[48] of a so-called Neanderthal man. Was he, it was asked, a victim of Noahian flood?

It was first brought to the notice of a scientific body in 1857 by Professor D. Schaafhausen who concluded that, despite some interesting characteristics, Neanderthal must be considered human and normal. In 1864 the German anatomist Mayer pointed out that pathology of the left arm indicated that Neanderthal had been afflicted with rickets, caused by lack of vitamin D. This in turn had caused the eyebrows to pucker causing brow-ridges. In 1872 Rudolf Virchow, father of modern pathology and anthropology, presented a closely reasoned paper demonstrating that the skull and limb-bones were not ancient at all, but from a man who had suffered from rickets (many Neanderthal children examined have shown rickets), arthritis in old age and several great blows on the head.

Thomas Huxley never saw the bones but recognised from reports that they were 'insensibly graded' from modern specimens, that is, were qualitatively human. Darwin never saw them either. Nor did William King, Professor of Anatomy at Queens College, Galway but this did not stop him placing the bones in a separate species, *Homo neanderthalensis.*

In 1886 a skeleton found in Belgium and associated with animal bones testified that the Hochdal find was not unique. Since then similar fragments have been found in China, Central and North Africa, Iraq, Czechoslovakia, Hungary, Greece and north-western Europe. Hundreds of bodies have yielded assorted fragments. It seems that Neanderthals were a race of men who suffered from malnutrition. They had prominent eye ridges, low forehead and an elongated brain case. Skulls found southerly are less Neanderthal in character than those found in the colder north.

In 1908 a Neanderthal burial was found at Le Moustier in south-east France. The classic remains were exhumed near La Chapelle-aux-Saintes, dated ~60 kyr and studied by Marcellin Boule from the National Museum of Natural History in Paris (and under whom a priest called Teilhard de Chardin studied). Despite the fact that brain volume was 1620 cm³ (as opposed to a large modern brain of 1450 cm³), he poorly reconstructed the specimen according to strong evolutionary preconception. Then, between 1908 and 1913 he proceeded to issue a series of scholarly papers on Neanderthal man, climaxed by a massive

---

[48] arm, leg and other fragments from several individuals.

monograph in three parts. *Ape-man was being delivered! The enormous influence of Boule's ape-like reconstruction lingers on. It represented a triumph for evolution over those who had believed the La Chapelle-aux-Saintes remains were of a diseased human.*

Thus our intelligent, beetle-browed, stocky relative was first portrayed as a forward-leaning, hunch-backed, morose-looking, crook-legged, shuffling cripple! Grafton Elliott Smith, anthropologist at University College, London, weighed in. He wrote about 'uncouth and repellent' Neanderthal Man whose nose was not sharply separated from the face, but was more like a snout. Stereotypes until recently drew a shuffling, shaggy hunchback with stupid gaze. Ape-man was being delivered in pictorial form! However, in 1997 the site of the original Feldhofer Grotto (lost due to mining) was rediscovered. So were more Neanderthal fragments along with fossilized pieces of a modern human

 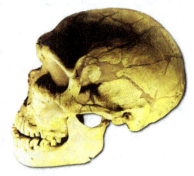

14.3 Neanderthal skulls; also very human reconstruction of *H. heidelbergensis* (see *fig.* 19.3) said to be the possible direct ancestor of both Neanderthal and yourself (*H. sapiens*). Indeed, today some classify this pair as *H. sapiens heidelbergensis* and *neanderthalensis* respectively.

(*Homo sapiens*) dated, as the original Neanderthal, at about 40 kyr. Co-existence? Evidence of stone tool production and fire used for cooking were also found. And now, after the discovery of about 500 specimens in Eurasia and the Middle East, it is recognized that no evidence whatsoever exists to substantiate Grafton Elliott Smith's fabrication. Thus the reader must judge, in such instances, where science leaves off and humbug begins.

Normal human brain size is 1250-1450 cm^3; Neanderthal's is 1400-1600 cm^3. If his brow is low, his brain is larger than modern man's. Learning this, Boule resorted to the Victorian pseudoscience of 'phrenology', a theory that the mental faculties are shown on the surface of the skull. He convinced himself that the frontal lobes were inferior in organization to those of modern man. Boule made mistakes yet not until 1957, with the throes of evolutionary parturition well over, did 'self-correction' hobble towards a different conclusion. This was revealed when two anatomists, W. Strauss from Johns-Hopkins University and A. Cave from St Bartholomew's Hospital medical college in London, took a second, closer look at the fossil from La Chapelle-aux-Saintes. Although the fossil was supposed to be typical Neanderthal, Strauss and Cave discovered that this particular person had suffered severe arthritis, affecting the structure of vertebrae and jaw. Boule should have detected this. The foot was not 'a prehensile organ', the neck vertebrae did not resemble those of a chimpanzee, nor was the pelvis ape-like. Boule had also arranged the foot-bones so that the big toe diverged from the other toes like an opposable thumb. This was the cause of the belief that Neanderthal had to walk on the outer part of his foot like an ape. His interpretation of the knee-joint, resulting in the 'bent-knee' gait, was also incorrect. Strauss and Cave punctured the myth he had created. They rumbled the prejudicial reconstruction and noted that Neanderthals walked just as humans do. They said: "There is....no valid reason for the assumption that the posture of Neanderthal Man....differed significantly from that of present-day man....if he could be reincarnated and placed in a New York subway - provided that he were bathed, shaved, and dressed in modern clothing - it is doubtful whether he would attract any more attention than some of its other denizens." Thus Virchow's diagnosis was vindicated - the bones from La Chapelle-aux-Saintes were from an osteo-arthritic geriatric with a skull of capacity 1600 cm^3. **But, from a propaganda point of view, the damage of Boule's faulty analysis had long been done. For fifty years portrayals of Neanderthal man always included a bent-legged stoop and 'primitive', stupid look. So the public mind became steeped in the notion that he was, after all, an inferior ape-man.**

The Victoria Institute noted that, in 1930, the skull and skeleton of a criminal executed in 1892 were exhumed in Australia.[49] The bones exhibited anthropoid ape characteristics, yet it was a fully modern man. Today the Arunta tribe of Australian Aborigines possess, as noted by Thomas Huxley, a large 'Neanderthaloid' skull and teeth, and some grow extra molars. In a generation, such 'savages' can take degrees at university and practise science. They are certainly as human as you or me.

If, in a hundred thousand years (102000 a.d.), planet earth is not a man-made desert into what sub-species will descendants cast our bones? Interestingly, fragments dated between 40 and 120 kyr have been discovered in the freezing Siberian cave of St. Denis. A couple of 'archaic' teeth (likened to those of *H. erectus*), a little finger (pinky) bone and 'difficult' fragments sum to represent *H. sapiens ssp. denisova*. *DNA* analysis (see particularly Chapter 22) further suggests congruence with Neanderthals, modern Melanesians and Australian Aborigines. In fact, warm it up! Join the party! It seems Denisovans, Neanderthals, archaic *Homo sapiens* and perhaps a Father Hominin (*H. erectus?*) interbred.[50] Few bones on so few sites excite much academic thesis and hypothesis but, new finds will likely reconfirm, Eurasian interracial (or interspecial if you take that line) interbreeding was the norm. Do humans tend to promiscuity or not?

Fragments of (mitochondrial) *DNA* from the de los Huesos ossuary in Spain (~400 kya) also match Denisovan; but degraded fragments of nuclear *DNA* (representing about 0.0005% of a whole genome) appear to show a stronger relationship with Neanderthals than Denisovans or modern humans. What does all this mean? Did humans evolve direct from *H. ensis*? Or from *neanderthalensis* who in turn had diverged (about 600 kya) with Denisovans from ancestral 'archaic' stock, that is, from *H. heidelbergensis* or similar? Complex speculation from a few bones of uncertain age, involving a proliferation of imaginary family trees, is staple Darwinian response. After all, if evolution is a fact man *must* have evolved somehow! So let's imagine how.

There is, however, no need for such industry within an archetypal frame. The situation looks, *prima facie*, as if we're dealing with variations (called sub-species or even species) within the limited

---

[49]   Transactions of the Victoria Institute, vol. 67, p. 21.

[50]   Nature 505: Four Make a Party (2-1-14 pps. 32-4).

   Nature 505: Complete Genome Sequence from Denisova (2-1-14 pps. 43-9).

   Nature 505: Mitochondrial gene sequence from Sima de los Huesos (4-3-14 pps. 403-6).

   Nature 507: Neanderthal in the Family (27-3-14 pps. 414-416).

   Science (AAAS) 11-9-2015: *DNA* from Neanderthal relative may shake up human family tree.

plasticity of human archetype (Chapter 9). *Allopatric and allochronological variation, genetic drift and founder effects occur in low-number, isolated populations. When any of these meet a positive 'interracial' mix occurs; and mutation negatively takes its toll.* We have lineage, of course, but not requiring evolutionary transformations from non-human types. **Is this interpretation of the facts invalid? Only philosophically!**

For example, it seems from mt-*DNA* studies that the Neanderthal population may have been about 25,000 individuals in total. Could Neanderthals be a product, within genus *Homo*, of above-mentioned variation? Perhaps, more likely, adaptive potential[51] plays the major role. Differing eco-morphological pressures such as severe ice-age conditions may naturally select certain predispositions; indeed, conditions might, with *DNA* epigenetically responsive to external conditions, actively promote in-built possibilities.

At any rate, *Homo sapiens ssp. neanderthalensis* had the same ear labyrinth and thereby balancing mechanism as ourselves. There is no intermediate between this and the ape-like pattern. And a typical Neanderthal differs but overlaps with modern man in brain-size, skull shape (broad, low and, perhaps due to enlarged cerebellum and motor cortex, slightly pointed at the rear), brow ridges, low forehead and rugged bones. His powerful jaw musculature might have contributed to skull shape. If, as seems most likely, this athletic race of men hunted skilfully in a harsh climatic regime then a short but large, thick-set and powerfully muscular body would, along with propensity for language that its hyoid bone[52] and large hypoglossal canal demonstrate, amount to a well-adapted survival package. Eskimos endorse this rationale. Certainly Neanderthals lacked a modern education but there is no reason to believe (other than to follow Darwinian theory) that they were less intelligent or talkative than us.

It is nowadays commonly mooted that Neanderthals became extinct about 40 kya after genocide by 'more human' types from 'out of Africa'. Neither side, it is asserted, previously crossed the Mediterranean (even at Gibraltar) because they had no boats so the 'more advanced' Africans must have radiated from the Middle East! Little or no intermixing occurred (though *DNA* speaks otherwise); Neanderthal women weren't enslaved by the hunter-gathering conquerors; and later Neolithic and Yamnaya farmers formed villages and set European civilization on course. This story omits to mention that Neanderthal remains dubbed 'too recent' have been found.

---

[51] see *SAS* Chapter 23 for fuller treatment of adaptive potential.

[52] found in Kebara Cave, Israel: Nature 328 (27-4-89 ps. 759-60).

Radiocarbon hits from the Shanidar cave in Iraq give 26 kyr; from Banyoles in Spain 17.6 kyr and from the Amud Cave in Israel from between 28 and 5.7 kyr depending on your dating method. We shall return later to Sima de los Huesos and El Sidrón in Northern Spain but meanwhile, since the readings don't fit evolutionary theory, 'contamination' takes the blame.

Do Aborigines or Eskimos actually live in caves? Why should roving hunters excavate and, as a few humans do today, build houses out of them? Nor do tents or huts survive ten thousand years. It is observed, however, that some modern troglodytes happily exist (as in Ürgüp in Turkey); that humans have often retreated to caves in times of crisis; and that, throughout human history, caves have been used as easily identifiable points to bury kith and kin (for example, Abraham buried Sarah in a cave and so laid to rest, in contemporary fashion, was Jesus Christ). It is more likely that use of caves was limited to short-term refuge or, since they would be easily recognisable, burial sites. And you may strongly argue that Neanderthals engaged in mortuary practice. Of over 120 sites at least 15 are burial (Shanidar Cave, Amud, Qafzeh (skull VI), Skhul (skulls IV and IX) and Tabun caves in Israel, caves at Le Moustier, La Ferrassie and La Chapelle-aux-Saintes in France, Krapina (Croatia), Sima de los Huesos and others. In short, over 50% of all 475+ Neanderthal individuals have been found in caves used as cemeteries.

In the 1960s a Neanderthal man was found buried on a bier of hyacinths, hollyhocks and other species identified from the pollen on the flowers and known for their medicinal properties. A ceremony, held maybe 60,000 years ago at a graveside in the Shanidar cave on the Iraqi side of the Zagros Mountains, seemed to clinch the matter. Unless your theory damns them brutes and you dream counter-plausibilities (e.g. wind or animals positioned the plants), Neanderthals were caring and cultured.

Outside the scrapyard what about this culture? Aren't stone and bone tools, hafted tools and use of fire crafts? Of what would you have made your tools prior to metallurgy? And stone bolas, such as could have been used in hunting game and certainly still are used in ranching, have been found. In fact, Neanderthal fossils have been found associated, like modern Cro-Magnon man and ultra-modern Holocene humans, with stone and bone tools (including 'hand-axes' that may have been used like a discus to hunt), artefacts, the controlled use of fire and mortuary practices. How much could remain of bows, boomerangs, catapults and other wooden instruments like boats? Neanderthals might, like Genghis Khan's men, have been horsemen (injuries on bones perhaps correlate with riding) but after only 800 years, how much sign of Mongol hordes remains? There is today a market for horsemeat but does that mean no equestrianism? Horse bones are often found at Neanderthal sites (not least Quinicieux in southern France). For these savages wild pigeon was a cooked delicacy. And would you expect Ice Age Man's dwellings, which might have ranged from igloo, yurt and tent to mud-wood-and-thatch cottages, to have survived intact, especially since a hunter life-style is nomadic? How much sign do Eskimos, red Indians and bush-men, in their culture, leave? Is an apparent absence of evidence evidence of absence?

Music and medication. There is hard evidence for music - a Slovenian flute with holes 'consistent with four notes of the diatonic scale' the same as modern flutes. Of course, since evolutionary Neanderthals were brutish ancestors, the flute attracts the usual plausible counter-claims. Maybe it is more advanced Cro-Magnon; or the holes were accidental punctures by a predator's teeth; or, if not a broken fraction, it is too short to have played musically. No early culture there of course!

And, as regards medication, an analysis of Neanderthal dental tartar has found evidence of herbal pharmacy (eg. pain-killing 'aspirin' from poplar bark).

Is hunting magic so inhuman? In a cave in Bavaria the skulls of thirty persons, with ornaments of deer-teeth and shells, had been placed in the earth. Charred remains nearby seemed to indicate cremation. The sun of day dies in the west and there, in later Arthurian legend, lay Avalon and the isles of the blest. Was it, therefore, symbolic that the skulls were sprinkled with red ochre and all faced, like the church door through which a coffin is carried, west towards that dying sun? Lacking modern technology what would your offerings consist of?

Hunting and even shamanic magic may have centred in a bear cult. At the Drachenloch cave in Austria, at an altitude of 8000 feet, a cubical chest of stones covered with a large stone slab was found. Inside were seven bear-skulls all with muzzles facing the cave entrance while six more were mounted in niches on the wall.

Pigments were, it seems, used as make-up, body-paint and for rock paintings such as found in Gorham's cave, Gibraltar. Did Neanderthals use jewelry or artistically fashionable clothes? Yes, their non-metallic bling was strung with claws, shells and perhaps less lasting elements. With awls why couldn't needle-eyes be bored and, with needles, sewing (perhaps with gut or fibres) could create a lasting fit. Intelligence does not require a thousand years to kit immediate needs and desires. They had skins, furs and stone blades so that a fashionable cut of suits, hats, hoods, gloves, simple shoes or boots and body wear must have been tailored, perhaps protecting modesty (as with a summer loincloth) or against a boreal ice-age (as with Eskimos).

As noted, prehistoric Neanderthal cemeteries have been found scattered throughout Europe and the Near East. The Sima de los Huesos, discovered in 1992 by Juan Luis Arsuaga at Atapuerca in Spain, is just such a site. It is deep, narrow and was never inhabited. Yet in this ossuary were deposited (not buried) on top of each other the remains of 28 or more individuals, with 3 finely preserved fossil skulls, dated at ~400 kya. There is such variation in the skulls, let alone the other bones, as encompasses the whole range of 'archaic' *H. sapiens* found in Europe. Indeed, as regards 15 cranial characteristics Arsuaga's fossils show 7 similarities with ourselves, 10 with Neanderthals and 7 with *H. erectus* (supposedly the precedent of Neanderthals) - suggesting that differences may be due to non-evolutionary causes occurring in a single, unique species, *H. sapiens*. It is fact that environmental conditions change (witness the oncoming of an ice age) and, due to genetic isolation, morphological changes can occur rapidly. Maybe Neanderthals *were* small, isolated populations of aboriginals both in the west and (witness skulls so far found in Indonesia, China and Australia), the east. In Europe most human-verging-Neanderthal morphologies (all but 4 of 130 individuals) are grouped in the central and eastern regions so that, yet again, interbreeding (called, in the jargon, 'genetic ingression') is suggested with a different race or races from the south (like those who may have produced the 400,000 year-old North African Tan-Tan figurine).

We'll look at the role of *DNA* in palaeontology later. For now, remember that the scientific definition of a species does not necessarily represent a viable breeding boundary. Fertile hybrids frequently occur. Thus, if you ask how Neanderthals did not survive, it might be answered that extreme climatic change or genocide played havoc with this race of men or, equally, if Neanderthals are *Homo sapiens neanderthalensis* that genetic ingression (Neanderthal with *H. erectus* as at Sima de los Huesos or Neanderthal with *H. sapiens sapiens*[53] in Israel and Europe) led to dilution of their particular characteristics.

Reconstruction of fragments found in 1978 in sea-facing Apidima caves on the Greek Peloponnese was reported by Nature (10-7-19). It transpires that of two crania one appears a morphological mixture of *Neanderthal/* modern *Sapiens* features and the other Neanderthal. The former, radiometrically dated (by U-Th method accurate to 250 kyr) at 210

---

[53]   witness humans and Neanderthals together in burial caves on Mt Carmel, at Magharet-et-Tabun and Mugharet-es-Skhul and in Galilee (Qafzeh).) This possibly occurred at Krapina in Croatia as well. Also, in 1997, the Feldhofer Cave in Hochdal was rediscovered. In it were both Neanderthal and human remains dated at 44 kya.

kyr makes it the oldest modern skull. It also implies of human/ Neanderthal coexistence for 170 kya! Certainly time to make at least good friends!

At any rate, an entire Neanderthal genome was first sequenced by Svante Pääbo of the Max Planck Institute for Evolutionary Anthropology and shows *DNA* between 99.7% the same as ours. Tests on several thousand Eurasians suggest that a large variety Neanderthal genes (about 15,000 or half their genome) are walking round today. By 2013 it had been suggested that, while an individual carries only 2-3%, over 20% of the whole human gene pool is Neanderthal. There is more than a trace of Neanderthal in you and me. Can't you thus reasonably infer Neanderthal and 'modern' women worked and gossiped while their children played? Then the men came home. What happens in mixed marriages? Some of each partners genes are 'bred out' but elements of both retained within the child. So that, if Neanderthals were a human variant could their genes not, over a possible 170,000 years of interbreeding, have been incorporated to such percentage as we find in current stock? It seems Alan Templeton correctly hypothesized that there is more genetic difference between contemporary humans than between us and Neanderthals. This, except for the nuance of biology textbook and magazine illustrators, makes them relatives of ours. Indeed, have you never spotted thickset characters with sloping forehead and with beetling brow as they spear seals, play sport, sit on committees or, closer home, whisk down your street to work? On average they would have, if 'Neanderthal', a bigger brain than you. Of course, you couldn't hold a conversation unless you learned her foreign language or she attended modern English class - but why should humans with a bigger brain and maybe intellectual capacity not speak? You might well presume they had the same archetypal instinct for rational order, symbol, grammar and the other aspects of communication. They might 'live rough' but equally could grace a hall of academe. Indeed, made-up facially according to Neanderthal skull form and ordinarily dressed, TV presenter Alan Titchmarsh put Strauss and Cave's pronouncement to the palaeoanthropological test in High Street. Did anybody bat an eyelid? No - well, except the odd appreciative glance from female non-Neanderthals! **The case of Boule shows how, with evolution, myths are born.**

In short, Neanderthals have cycled from first diagnosis as a 'flood victim' through the fervent mythology of Victorian imagination to a cooler Elizabethan analysis. In an evolutionary frame this restores them, *H. sapiens neanderthalis*, to a relatively 'primitive' ancestor - neither dim nor lumbering. Their bodies were well-built to live a hard, hunter-gather life. Neanderthal genes exist in you and me; so-called Neanderthals and *Homo sapiens* must have interbred. **In a non-evolutionary frame simply delete 'primitive', insert 'eco-morphologically varied' and know him and her, within the human family, as one of us.**

15. *Homo erectus/ Java Man*

Dubois

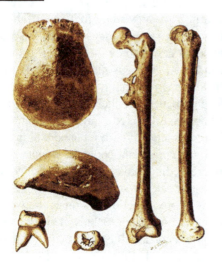

15.1 Three-bone composite (two angles each) of Eugene's
Dumb Caliban *aka* Java Man; also figment of aforesaid
and suckling wife by artist Gabriel von Max
commissioned for Haeckel's 60th birthday in 1894.

The search for a missing link began in earnest when Ernst Haeckel,
a German philosopher and writer, entered the scene. Haeckel
hypothesized an ape-man and blessed it with the name *Pithecanthropus
alalus* - the speechless ape-man. He thought it might be found in
Southern Asia, or maybe Africa. By 1887 the imagination had fired a
pupil to sail for the East Indies. Eugene Dubois was keen, even
determined, to hunt for and find this long-awaited missing-link.

In Sumatra Dubois found little. But in 1888, hearing that a fossil
skull had been found in neighbouring Java, he sailed over and obtained
it; then he found another fragment on its site of discovery, a place called
Wadjak. A lack of precision plagues the geological details of this site
where, in 1890, Eugene Dubois found a further human skull (Wadjak II)
and fragments of a skeleton. The brain volumes of men and women can
vary from ~790 to over 2000 cm³. Apes vary from about 90-720+ cm³,
according to body size. The Wadjak skulls were fossilized but, being
1550-1650 cm³ in capacity, were too like modern man to be claimed as
missing links. So Dubois bided his time. He stowed away the skulls both
in Java and later Holland until the early 1920's.

In 1890 at Kedung Brabus he found a jaw containing the root of a tooth. In September 1891 at Trinil by the Solo river his Malay coolies unearthed a large molar tooth and a month later the fossil skull-cap of an ape-like creature - now famous - was discovered. In August of the following year he found a human femur (also completely fossilized) about fifty feet away which, despite the distance, he imagined was related to the beetle-browed top section of skull. He also found another molar about ten feet from where the skull-cap had been found; and several more human femurs were dug up.

After correspondence with Haeckel, he ignored the human skulls from Wadjak and the Trinil femurs, except one. This latter, he decided, belonged to the owner of the skull-cap; he declared them both to belong to a creature which seemed admirably suited to the role of the missing link. **Haeckel, without seeing the evidence, immediately telegraphed back: 'From the inventor of *Pithecanthropus* to his happy discoverer.' Thus was the first, upright-walking ape, *Pithecanthropus erectus aka* Java Man, unearthed and named.**

At this point it is worth interjecting that fossilization of organic material can, in the right circumstance such as heavily mineralized water, become rapidly fossilized. Japanese scientists[54] reported that wood suspended in warm volcanic water at Tateyama volcano had been 40% mineralized in 7 years. Laboratory increase in temperature and pressure (as might well also be found in an area of geological upheaval) can further increase both percentage fossilisation and the rate. The area of Javan finds is, of course, heavily volcanised.

Secondly, if the skullcap is *erectus* and the femur human, then you might presume the two species coexisted; if, on the other hand, they belong together then what is the difference between *erectus* and *sapiens*? They'd be the same!

Thirdly, at this time Dubois was with medical but without geological training. Mapping and measuring were not his strong suit. He worked without precision, photographic or otherwise. His was more a theoretical than physical dead reckoning. In fact, he employed coolies to dig without his supervision and his only personal find *in situ* seems to have been at Wadjak (skull II). His standard of excavation and documentation would be unacceptable today.

So, in confusion and secrecy, the myth of Java Man was propagated - invented at just the right time to clinch the evolutionist case. Von Koenigswald, who excavated in Java from 1931-40, notes in *Meeting Prehistoric Man*: "When Dubois issued his first

---

[54] Sedimentary Geology 169 (July 2004 ps. 219-228).

description of the fossil Javanese fauna he designated it Pleistocene. But no sooner had he discovered *Pithecanthropus* than the fauna had suddenly become Tertiary."[55] Actually the surrounding flora was found to be contemporary and a date of 10 kya could well have been argued. But Dubois wanted his 'man' to be early - as close to the true ancestral stock as could be managed. So, continues Von Koenigswald, "...he did everything in his power to diminish the Pleistocene character of the fauna. The criterion was no longer to be the fauna as a whole but only his *Pithecanthropus*."

Between 1889 and 1892 Dubois did file what were probably contractual reports mentioning the Wadjak excavations with Javanese official bodies. Maybe no bureaucrat read or understood them. They were certainly not made 'public'. Indeed, in 1894 he published a paper about *Pithecanthropus erectus* which made no mention at all of the Wadjak skulls. That would have cast doubts on the simian-seeming Trinil skull-cap, whose capacity he had somehow estimated (at 850-900 cm³) as set between man and ape! So did Dubois stow the human Wadjak skulls under his floor-boards for over 20 years to prevent any deflection of glory from *Pithecanthropus*? Because *Pithecanthropus* came, as palaeontologist Gustav von Koenigswald later wrote (in 1956), **"...just at the right moment at a time when the conflict around Darwinism was at its height. For the scientific world it constituted the first concrete proof that man is subject not only to biological but palaeontological laws"**[56] *Beats on the evolutionary drum were flowing faster, louder - each presumed a death knell for 'unscientific' theories of design.*

However, in 1895 anatomist and anthropologist Sir Arthur Keith FRCS decreed Java Man human. And when, also in 1895, he exhibited the skull-cap and thigh-bone in Berlin Virchow (who had no time for his former pupil, Haeckel) refused to chair the meeting, saying that in his opinion the bones represented a giant gibbon, not early man at all. The thigh-bone, he said, had not the slightest connection with the skull.

In 1900, after a period of actively promoting his ape-man, Dubois 'went quiet'. Then, in 1918, Australian anthropologist, Professor Smith formally reported on the first Australasian.[57] This seems to have

---

[55] Von Koenigswald G.H.R. *Meeting Prehistoric Man,* Thames and Hudson, 1956, p38.

[56] Von Koenigswald G.H.R. *Meeting Prehistoric Man,* p. 26.

[57] the skull of Talgai man, found in 1886 about 80 miles south of Brisbane, was human but his palate and teeth seemed a little ape-like. Was he, it was asked, an aborigine, perhaps one from Tasmania? Later we shall check other bones found in SE Australia.

provoked a jealous Dubois to show the Wadjak skulls; and in 1924 he gave details of the jaw found at Kedung Brabus in 1890, which in 1891 he had called human but now identified as part of *Pithecanthropus*. Also in 1924 he published a long-delayed but definitive paper on the Trinil skull-cap followed, in 1926, by one on the femur. Only in the early 1930s were other human femurs announced. The earlier suppression of human material was calculated to highlight the dubious 'ape-man' bones. The irony is that, although Dubois himself had by 1936 said he thought *Pithecanthropus* might be a *large gibbon*, photographs of painted models of slouching 'Java Man' were being reproduced in children's and higher level textbooks. They still are, despite the criticism of Boule and Vallois that such models are 'pure flights of fancy'. **In the popular and scientific press *Pithecanthropus*, beginning as a dream of Haeckel and realized by skullduggery, engendered its own growth industry and that continued and continues to thrive.**

The Selenka Trinil Expedition

Act II in the Javanese theatre involved restraining counterpoint and has, therefore, been 'suppressed' from wide, textbook assimilation.

In 1907-8 Professor Lenore Selenka led an expedition which, far more extensive and scientific in its scope than any of Dubois, used over 70 coolies to remove 1000 cubic metres of earth from exactly the same location on the Solo River. However, the scientists concluded that Trinil's volcanic, flood-deposited sediments were too young to throw light on evolutionary origins. For example, the flora was modern and so were most of the gastropod forms. Although forty-three boxes of fossils were dispatched to Europe, no confirmatory trace of *Pithecanthropus* (Java Man) was found. The researchers did, however, dig up charcoal, hearth foundations, human bones and artefacts from the stratum. An excellent report was compiled (*Die Pithecanthropus-Schichen auf Java*) which, while wishing to confirm an earlier evolutionary perspective, actually cast real doubt on it. Frau Selenka believed *Pithecanthropus* was an aberrant contemporary. For this reason she apologised to co-worker Max Blankenhorn for lack of evidence supporting Dubois' hypothesis. The expedition, although precisely scientific in its methodology, did not 'produce the goods'. Thus he and others, lacking objectivity due to their evolutionary point of view, called it 'fruitless' and 'disappointing'. Maybe it is also the reason why her work has been and still is widely ignored.

Her report, published in Leipzig in 1911, was never translated into English. However, Keith gave a short account in Nature[58]. He noted

---

[58] Nature: vol. 87 no.2176 13-7-1911; The Problems of Pithecanthropus.

that the area was one of considerable volcanic activity (nearby Mount Lawu had, for example, exploded in 1864 and 1875; there had been relatively recent change in the flow of the Solo river, said by Dubois to have separated his *Pithecanthropus* skullcap and human 'upright-walking' femur; and there was argument over dating but Selenka concurred with young (perhaps Holocene) geology. Later, in 1968, Kenneth Oakley of Piltdown fame correlated Trinil river gravel with gravel containing similar fossil fauna but found elsewhere in Java. He thus estimated the age, by potassium/ argon dating, at around 500 kyr. A 'fruitful' correlation.

von Koenigswald's Research

Act III. In 1931 Gustav H.R. von Koenigswald, sent by the Geological Survey of the Netherlands East Indies, arrives at Bandung. Some human-like material, called the Solo or Ngandong skulls, are discovered about 40 miles from Trinil. Artefacts are found nearby by which the skulls are designated Neanderthal.

15.2   Teilhard de Chardin (1881-1955)

In retrospect wisely or unwisely, 1935 von Koenigswald invited Jesuit priest and ardent evolutionist Teilhard de Chardin - see also Piltdown - from China in order to establish a palaeontological connection between Javan and Chinese fossils. Although up to that time none had been discovered, it seems that he and von Koenigswald entered a cave in the Patjitan (Pacitan) area and found what proof they needed *on the floor*. What luck! An abundance of isolated teeth - orang-utan, large gibbon, bear and so forth - were 'absurdly' similar to the

88

fossil-bearing deposits of Kwangsi. For the first time orang, gibbon and bear were found in Java. A correlation was established, it seemed, with southern China. And it was hoped the teeth on the floor of the cave would be seen to link Java with Peking man (see below), already discovered 3000 miles away. Teilhard also hoped that the discovery of *Stegodon* (elephant) fossils in the Karst lands of Gunung Sewu (Pacitan) would establish a link between the Solo basin of Java and the Kwangsi fissures of China. Shortly afterwards von Koenigswald found just what he needed. We are ignorant whether the fossil's radio-activity was commensurate with that of Ichkeul fossils. Nevertheless, another plank was added to Teilhard's 'theory of the origin of man'.

In 1936 von Koenigswald failed to log the exact site of his Modjokerto child. This was the skullcap of a child found near Surabaya. He recognized its humanity but called it *Pithecanthropus modjokertensis*. If it was human, however, Dubois objected. Von Koenigswald acceded and re-named it *Homo modjokerto*. **It was dated at ~1 mya but, in the 1990's dating specialist Garniss Curtis (who with Getty support founded the Berkeley Geochronology Centre) corrected this to 1.81 mya.** In 2006 detective work led to a precise identification of the site that led to a reduction of date to ~1.5 - still very old for an evolved *H. sapiens* or *erectus*! Here matters lie.

In 1937, having lost his Geological Survey Grant, von Koenigswald was introduced to an American donor, the Carnegie Institute, Washington. No records of a meeting in Philadelphia exist but it seems he persuaded his benefactors that Sangiran was about to yield 'proof' of an ape-man. From 1935-40 he explored this area (further west along the Solo River) with the help of natives who were offered a reward for every piece they found. As well as this temptation for poor natives to plant bones inadequate supervision, due to his station at Bandung being over 200 miles from Trinil, left further scope for the same type of malleable imprecision found in the work of Dubois. He found fragments of jaw-bones, teeth, skulls and a skull-cap but no limb-bones; these pieces were designated *Pithecanthropus* II, III and IV.

Pithecanthropus II (1937) was retrieved as 40 fragments because, having offered an inflated award per piece, the natives had broken down what they found. The 'almost complete' reconstruction only incorporated 30. The ear was human but, when contacted, Dubois calculated from photographs that the skull had been improperly rebuilt. He accused von Koenigswald of fakery but, when the latter rejected this as nonsense, relented. However, various missing parts made measurement of cranial capacity impossible.

In 1938 famed anatomist and anthropologist Franz Weidenreich reported from China that the Modjokerto skull and *Pithecanthropus* II showed that Dubois' *Pithecanthropus* was human-like. In this year also von Koenigswald designated three pieces of a juvenile skull *Pithecanthropus III*.[59] Since these seemed similar to ones found in China he arranged to visit Weidenreich for a comparison. Dr. Davidson Black - agreed but thought his 'Peking Man' (Chapter 17) was our true ancestor.

Pithecanthropus IV (or *Pithecanthropus modjokertensis*) is composed of a jawbone and unconnected rear part of a skull (the latter requested from its collector at a later date). When placed with Mandible 'B' (from Sangiran 1935, which von Keonigswald had labelled Pithecine and Dubois human) within the outline of a gorilla's skull the pair fit well.

While Boule and Vallois[60] both report on the ape-like nature of these fragments others, such as le Gros Clark, were very critical of the considerable confusion which surrounded their reconstruction and classification (which stands at *Homo erectus*). However, in 1996 Garniss Curtis dated the Ngandong/ Solo *erectus* fossils at between 27 and 53 kyr. Such dates would indicate that Javan *erecti*, much younger than their African *erectus*/ *ergaster* counterparts, must have co-existed with modern Cro-Magnon *Homo sapiens*. This why Curtis' dates rocked a few foundational assumptions! And perhaps why in 2011 SoRT (the Solo River Terrace Project), using argon-argon and other dating metrics, came up with a somewhat 'safer' age, one definitely between 550 and 143 mya.

Certainly Dubois seriously misinterpreted his finds and dating, at 500 kya as required by theory to close the ape-man gap, is suspect. Three separate explorations found only fragmentary evidence subject, on occasion, to whimsical interpretations. And many less agendum-driven anthropologists of the time thought Java Man was human. **If so, this version of man designated *Homo erectus* was something like a small Neanderthal. Is it, therefore, possible that *H. erectus* is, like *H. sapiens neanderthalensis*, really *H. sapiens erectus*?**

---

[59] Nature: Vol 141 26-2-38 p. 378.
[60] in the book Fossil Men.

16. Piltdown

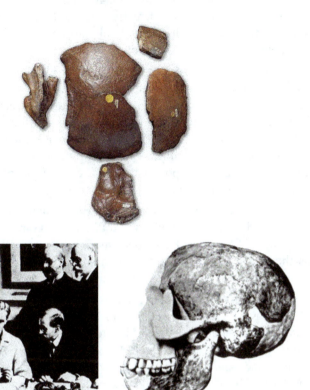

16.1   Piltdown fragments + 7 wise men + their object of veneration.

'Archaeological forgery is a tangible form of historical fiction' (Ian Haywood: Faking It). And for those convinced of evolutionary creation, grist to their propaganda mill did not long follow Darwin's book of 1859. In 1863, at Moulin Quignon near Abbeville in France evolutionist and archaeologist Boucher de Perthes offered a large sum to any labourer who dug up a human remain at the spot where he had previously found flint tools. A tooth and later the lower half of a jaw emerged. However, the knapped tools of a dodgy dealer, 'Flint Jack', had flooded the black market for archaeological fakes; even if de Perthes' *were* genuine no-one could actually tell newly-knapped from old (even when Flint Jack owned up to his trade). But analysis definitely found the jaw modern.

The passion for finding what you know that evolution must have left did not, perhaps, propel de Perthes to intentionally fraudulent behaviour. But if his and, more so, the story of Java Man were unsavoury, that of the Piltdown forgery is an intrigue worthy of a 'whodunnit?' - a tangled web of spurious evidence, innuendo and suspicion. It will surely rank as one of the most notorious, intentional scientific frauds of all time. Three recent, reasonably brief detective accounts each 'point the finger' slightly differently. Where this book can no more than summarize the main events, they are worth a closer look.

Charles Dawson (*16.1* back row 3rd from left), a Sussex lawyer, was well known locally as a collector of fossils. In the 1880s he presented a collection of fossil reptile bones to the British Museum and struck up a lifelong friendship with the austere keeper - Arthur Smith Woodward (*16.1* back row 4th from left). This self-made man, unpopular at the museum due to his aloofness, was an expert on fossil fish bones. He was certainly not an anthropologist; there were few people in Britain at this time who could fill that role and he was not one.

Edwardian pride took a knock in 1907 when a very large jaw with small, modern-looking teeth was found in the Gunz-Mindel interglacial deposits of Germany. Heidelberg man, as it was called, was identified as the oldest fossil remains of man so far discovered. The search for similar fossils elsewhere in Europe intensified.

In the following year Charles Dawson claimed that he had unearthed a British saving grace - several fragments of thick fossilized skull, stained with iron for their full thickness, from a gravel pit at Piltdown, about halfway between Uckfield and Haywards Heath. In 1911 he claimed to have found several more skull fragments (which matched the first), together with other fossils and artefacts from the site. Early in 1912 he informed Smith Woodward at the Museum that he could equal Heidelberg man. In consequence, from May 1912 these two went several times to Piltdown, accompanied from time to time by other members of the British Museum staff. Another who occasionally worked with them was a young French geologist we've already met (*fig*, 15.3), Pierre Teilhard de Chardin, the Jesuit priest with a passionate affinity for fossils and a renegade interest in evolution - including the evolution of man.

Teilhard's childhood had been spent in the Auvergne mountains of France. From his mother, who was a great-niece of Voltaire, he learnt spiritual intensity; and from his surroundings, love of the 'permanent' rocks. When the Jesuit Order was expelled from France to Jersey he was, as a novitiate, so entranced by the fossils he found that he began to doubt his priestly vocation. His fears that palaeontology might conflict with Christianity seem to have been assuaged and he was sent from 1906-8 to lecture in physics at Cairo University. Here he is almost certain to have heard of any interesting sites in North Africa. It is reported[61] that he

---

[61] Millar R. The Piltdown Men, Gollancz, 1972 p. 232.

stayed at Ichkeul near Bizerta in North Tunisia, a site where *Stegodon* (elephant) fossils are plentiful.

In 1908 he came to Ore Place seminary in Hastings, where he was ordained. Again the idea of evolution, which became a master sentiment, began to work in him and he began to search for fossils. In one of the local quarries, on 31 May 1909, he met and befriended Charles Dawson.

Now, three years later, Dawson, Smith Woodward and Teilhard unearthed more skull fragments and other fossils. One day Dawson, in the absence of Teilhard but before Smith Woodward's very eyes, hit gold in the form of an ape-like jaw-bone. It was unfortunately broken in two places, at the hinges and on the point of the jaw, which made it difficult to establish any clear relationship with the rest of the skull that Dawson was assembling from fragments. The canine teeth, which would have helped to identify it as ape or man, were also, unfortunately, missing.

16.2   Imaginary Piltdown on the Prowl.

The finds were announced before the Geological Society in London on 16 December 1912. *Eoanthropus dawsonii* (Dawson's Dawn man) was reported as the British answer to the German jaw, and toasted in a thousand sophisticated drawing-rooms. Was not the first man British? Not only Britain's kudos, but its discoverer's too, was considerably enhanced. On 29 August 1913 Teilhard stayed overnight with Dawson and went next day with him and Woodward to the Piltdown pit. Lo! There appeared one of the two missing canine teeth. Arthur Smith Woodward reported that they excavated a deep trench in which Father Teilhard was especially energetic. When he exclaimed that he had picked up a canine tooth, the others were incredulous, telling him that they had already seen bits of iron-stone that looked like teeth on the spot where he stood, but Teilhard insisted that he was not deceived. They left their digging to verify his discovery. There could be no doubt about it - Teilhard had found a canine from the previously discovered jaw.

The tooth was pointed like an ape's but worn in ways that suggested human origin. It closely resembled the canine in an imaginative reconstruction that Smith Woodward had caused to be created for the 1912 Geological Society meeting. It was the decisive missing clue and was reported, along with other animal fossils, including a third piece of stegodon tooth, at the next (1913) meeting. Interestingly, the stegodon tooth was found to contain 0.1 per cent uranium oxide and to have a particularly high level of radioactivity, unusual for European fossils but found in those from Ichkeul, Tunisia. The Ichkeul site was not publicly identified until 1918, so that it is very unlikely the Englishmen had visited it. But Teilhard had.

Although the vast majority of scientists eagerly, uncritically accepted this third strong beat on the evolutionary drum not everyone was impressed by the Piltdown finds. There existed considerable suspicion in the minds of two amateur palaeontologists from Sussex, Captain St. Barbe and Major Marriott. They had, on separate occasions, surprised Dawson in his office staining bones and did not trust him. On 13 November came an independent opinion that doubted the find. *Nature* published a letter from David Waterston of King's College, London, which ended: "It seems to me to be as inconsequent to refer the mandible and the cranium to the same individual as it would be to articulate a chimpanzee foot with the bones of an essentially human thigh and leg."

But in 1915 a postcard from Dawson to Smith Woodward proclaimed more fossils from a second Piltdown site. This site is not identifiable today. The cranial bones were more fragments from the skull that was found in the first pit. Then in 1916 Dawson died, and in the following year Smith Woodward published the 'confirmatory' Piltdown II finds. And across the pond in 1921 Henry Osborn, President of the American Museum of Natural History, declared that skull and jaw unquestionably belonged together.

However, in 1925 and again in 1951 a member of the Geological Survey, Frederic Edmonds, severely criticized the geology arraigned to assign an age to the fossils. The palaeoanthropologists, of course, ignored him. As late as 1948 Sir Arthur Keith (*16.1* front row centre) penned a Foreward for the late Sir Arthur Smith Woodward's eulogy to Piltdown Man called 'The Earliest Englishman'. But The British Museum authorities, in a manner not designed to allay suspicions, now kept the Piltdown fossils under lock and key. Even such anthropologists as Louis Leakey were only allowed to handle the casts (from which any fraud could not be detected).

Only in 1949, when a method for determining the ages of fossil bones by analyzing their fluorine content was revived, were critical tests carried out on the Piltdown fossils. The human skull (carbon-14 dated to about 600 years old and probably from a local black-death grave) and ape-like jaw (almost certainly an orang-utan jaw from Java, Borneo or Sumatra) contradicted updated ideas about man's evolution; thus they aroused suspicion. Both pieces were much younger than the other fossils (such as Teilhard's stegodon tooth) and artefacts with which they were supposed to have been found. Most specimens had been artificially stained with bichromate. The canine tooth had been filed, coloured and packed with grains of sand. An elephant bone associated with the skull, of a type found also in the Dordogne and Egypt, had been cut with steel instruments into an improbable 'bat' shape. Therefore when in 1953 Dr Kenneth Oakley, in collaboration with J.S. Weiner and W. le Gros Clark, completed investigations and unmasked the fraud, concern was such that a motion was tabled in the House of Commons "that the House has no confidence in the Trustees of the British Museum....because of the tardiness of their discovery that the skull of the Piltdown man is a partial fake." After all, the Nature Conservancy had (in 1950) just spent a lot of taxpayers' money tidying up the Piltdown site, which had then been declared a National Monument (as it turns out, to Fraud).

So now, long after propaganda value had been richly secured, did the British Natural History Museum at last issue notice that the Piltdown 'skull' was an intentional fraud. The faking of the mandible and canine was skilful: it was certainly a determined and unscrupulous hoax. Sir Solly Zuckerman, who later reviewed the story, considered the hoaxer knew more about primate anatomy than the experts who were several times deluded.

Whodunnit? No doubt, the BMNH (British Museum of Natural History) strongly promotes Darwin's theory of evolution but is perhaps unlikely, even at a time of evolutionary frenzy, to have conspired in fraudulent promotion. However, it certainly took, despite earlier intimations of possible skullduggery, forty years before it was decided the time was ripe for public admission. If there was fraud, there were fraudsters and, it seems, gullible 'experts'. Thus, since 1953 fingers of blame have been variously pointed. BNMH has consistently deflected those aimed its way. It has tended, since Joe Weiner made a couple of visits to Piltdown in 1953, to sway towards Dawson as the main perpetrator.

Did Dawson know enough of primate anatomy to fake well enough? Numerous people have researched the murky entanglement of deceit. However, at least Zuckerman, Beverley Halstead and Malcolm Bowden would implicate Martin Hinton (BMNH) as a collaborator in the hoax. After all, the latter later retracted a 1926 endorsement of the fossils saying that he would have diagnosed the finds as fraudulent if he had been allowed to inspect them more carefully! Halstead believed he supplied the medieval orang-utan jaw. And later a bag was found later in a BMNH loft with his initials and incriminating material in the form of scalpels with animal bones and teeth carved and stained in a manner similar to the Piltdown finds. Several factors enhance the suspicion of guilt but, in this case, could the motive for chicanery simply have been to make a fool of his boss, Sir Arthur Smith Woodward? Or was it, in reality, to boost public acceptance of the theory of evolution and the contemporary 'expert' notion of man's origin in Europe? Grafton Elliott Smith (*16.1* back row 2nd from left) had foreseen just such a find.

What role in this plot did the hovering Teilhard de Chardin play? It was almost certainly 'evolution-mad' Teilhard who planted the canine tooth. And in 1954 he gave a most evasive interview to Weiner and Oakley. It is very likely he was part of an original group of conspirators (as fraudsters are). But, for whatever reason, the establishment tends to protect this 'scientific' evangelist who pops up across the early twentieth-century palaeoanthropological scene.

For further clues start with the authors that this footnote recommends.[62] Piltdown is a case, if ever there was one, which illustrates the imposition, by 'scientific' evolutionists, of strong hope, desire and prejudice on a few bones. There may be more of its false trail to be unearthed.

---

[62] Harrison Matthews L. '*Piltdown Man*' (a detective story in ten parts), New Scientist, 30 April-2 July 1981.

Bowden M. '*Ape-Man, Fact or Fallacy?*' and '*Science vs. Evolution*' Sovereign Publications, Bromley, 1981.

Gould S.J. *The Panda's Thumb,* W. W. Norton & Company, 1980.

Weiner J . S. (with. Stringer C.) *The Piltdown Forgery,* OUP 2003.

17. *Homo erectus/ Peking Man*

17.1   Oriental Skullcap (*Sinanthropus* II).

Peking man? You've had a taste of all the wrangling. Be prepared for more. In practice macro-evolution's not as clear-cut as insistence that it must, somehow, have happened. So what's the fuss about a few frauds, farce or fallacies if you have faith the underlying process is completely true?

In 1903 Professor Schlosser of Germany had, on examining a number of fossils purchased from a druggist's shop in China, found a tooth he considered anthropoid and suggested that early man might be found on that continent. The tooth has since disappeared but it may have served to spark interest. It seems that, as early as 1912, Father Licent (a Jesuit and museum director) had directed attention towards Choukoutien thirty-seven miles from Peking. Choukoutien (or Zhoukoudian) means Dragon Bone Hill, where the Chinese for fossil is 'dragon bone'.

In 1922 two human molar teeth were found. On their basis Davidson Black (then Professor of Anatomy at Peking Union Medical College) made extravagant claims about 'primitive man'. Thus, though molar, they had bite because in 1926 Davidson Black sunk them into the Rockefeller Foundation and incisively extracted finance for a couple of palaeoanthropological explorations. Black was a Canadian who had studied with Grafton Elliott Smith in 1914 while the latter was researching Piltdown Man. Do you remember the single-toothed Hesperopithecus affair? **On the basis of a single tooth found on the 1926 Rockefeller exploration Black, a believer that mankind had its origins in the east, announced *Sinanthropus pekinensis* (Peking Man).** This marks the start of a complex, unsatisfactory affair upon which, as in the Piltdown case, the last word has not been said.

However, as regards results and grants, it had been enough. From 1929 the Rockefeller Foundation granted $20,000 per year to carry out excavations. It further agreed to finance a Cenozoic Laboratory with Davidson Black as Director. **Thus Black became, as opposed to Dubois with Java Man, the promoter of Peking Man.** And Teilhard de Chardin, whose desire to promote evolution by whatever dubious means did not end with Java or Piltdown men, became advisor and collaborator with special reference to geological and dating concerns.

At this point Teilhard had been 'banished' on a mission to China by the Jesuits, not for fraud (which remained unsuspected) but simply because of his desire to promote evolution. From 1923-46, as a consultant to the National Geographical Survey in China, he travelled the land extensively. But he was based at Tientsin with Father Licent in his Geological and Botanical Museum where, in 1942, he lectured Chinese students/ future communists on how they had evolved from animals and in 1943 had the lecture printed in pamphlet form.

By 1928, 575 boxes of bones had been sent from limestone Dragon Bone Hill to Peking. In all, about thirty skull-bones, eleven jaws and 147 teeth of the 'ape-man' were recovered. In only five cases were sufficient bones of the cranium available to allow a skull to be reconstructed adequately for measurement which, each time, turned out around the 'correct' litre of mind, that is, the 1000 cm³ mark that hangs between average human and large simian brain size. The fragments were found in an enormous (seven-metre high) heap of ashes. They were mixed up with animal remains, mostly deer and other edibles. There was no difference between bones found at the top and bottom of the deposit i.e. no evolution was apparent.

Pei Wenzhong, also of the Cenozoic Laboratory, honoured by the evolutionary-minded Communist Party and sometimes known as 'the father of Chinese anthropology', was in charge of the 1929 excavations. He found a skull that lacked base, jaw and front of face. This, following Black's reconstruction, was chosen to head the *Sinanthropus* thrust. Of it Marcellin Boule wrote:

"Black, who had felt justified in forging the term *Sinanthropus* to designate one tooth, was naturally concerned to legitimize this creation when he had to describe a skull-cap."[63]

It was not until 1931, however, after the visit of his former tutor Elliot Smith that Davidson Black issued the model and a long article for which he was duly elected a Fellow of the Royal Society. His official description differed, however, from an eager, prior dispatch by Teilhard, published in the July 1930 issue of the French *Revue des Questions Scientifiques*. In this dispatch he notes a surprisingly small

---

[63] Boule M. and Vallois F. *Fossil Men,* Dryden Press, 1957, p. 141.

brain capacity and ape-like appearance. Grafton Smith corroborated his assessment. The face below the eye-sockets is missing. In Black's model, though, only the skull-cap, from which it was not possible to check the alleged brain capacity, was present. This item closely resembled Java Man. It was the reconstruction on which Davidson Black had worked long, secretively and hard, by night. This reconstruction is the only evidence that its brain capacity was of an ape-man's litre.

A first-hand report by Abbé Breuil of his 1931 visit to the site demonstrated that there had existed at Choukoutien an industry of a nature far too large and advanced for it to be attributed to the small-skulled animals called *Sinanthropus*. Perhaps it had serviced a city on the site of the present Beijing. Thousands of chipped quartz stones had been transported several miles and used as tools. Enormous furnaces had been kept burning for long periods, leaving the large deposits of ash. Bones had been worked and cut to a level easily comparable with that of Neanderthal man. Breuil also mentioned 'bolas' stones, used for entwining the legs of creatures, which it would have taken considerable ingenuity to conceive and then construct. However, Mr. Pei could not provide locations for the bones that Breuil studied from Choukoutien and later announced their 'unfortunate' disappearance!

Boule was also invited to the site but when he saw that the only evidence provided was battered monkeys' skulls, each with a hole in the top, he was vexed, denounced Teilhard and ridiculed the idea that the owners of the skulls could have carried out the large-scale industry revealed. He wrote: "We may therefore ask ourselves whether or not it is overbold to consider *Sinanthropus* the monarch of Choukoutien when he appears in its deposit only in the guise of a mere hunter's prey, on a par with the animals by which he is accompanied."[64]

Another Jesuit who was in China at the time, Rev. Patrick O'Connell, also denounced Peking man. Choukoutien was, he believed, quarried; lime-burning was carried out there a few thousand years ago. In the course of time the hill was undermined and a landslide occurred which covered everything with thousands of tons of debris. With the aid of the Rockerfeller grant this debris was removed and the remains found. Fossil skeletons of baboons and macaque monkeys, which do not differ from contemporary forms except for their greater size, had been found in the district. He believed that these large monkeys, often with neat holes drilled in their skulls, were captured for food. Their meat would be too tough but the heads would be brought back for the

---

[64] *ibid.* (as footnote 57) p. 145.

brains which, when cooked, would be a delicacy. This would explain the holes through which grey matter could be spooned: also the fact that only *Sinanthropus* skulls and jaws but no post-cranial materials have been discovered; and their situation in the ash after being thrown on the fire with other culinary garbage.

Speculative disagreements, always within the evolutionary frame, characterize the nature of palaeoanthropology. Teilhard (with Mr. Pei) denounced the view of O'Connell, Breuil and Boule. He did not agree that this was a large industrial site where large monkeys had been brought, decapitated (virtually no post-cranial parts are found) and butchered for their cerebral delicacy - as still happens in parts of China and Africa. The pair produced a lengthy write-up (Fossil Man in China, 1933) which plays down, to the point of effectively suppressing, Breuil's mention of the large ash deposit and his identification of an advanced tool-making and industrial site with extensive use of fire. Why? It might seem to humanize *Sinanthropus* whose idea Teilhard all along championed. Of course, the Cenozoic Laboratory's insistent aim, its *raison d'être*, was the evolutionary establishment of ape-man. Thus its wishful interpretation of the evidence from which, however, in 1934 Teilhard was obliged to grudgingly and partially withdraw.

In 1933 Mr. Pei announced a few skeletal remains (including three fragmented skulls) of six humans in an 'upper cave', an S-shaped vertical recess probably formed when a landslide occurred. The skulls were fossilized in an already smashed condition. On 15 March 1934 Dr. Davidson Black, on entering his laboratory in Peking to examine these fragments, fell down dead among them.

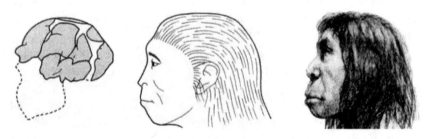

17.2   Nellie (by Mrs. Swann)
and her husband (from The Sphere Newspaper).

His successor, Professor Franz Weidenreich, took charge from 1934-40 and worked painstakingly on re-casting Black's reconstructions. The skulls, except skull X at 1225 cm³, would again turn out as 'standard ape-man litre'. A second model of *Sinanthropus* using Skull XI fragments used plaster to fill its best-guess spaces.

Facial bones were 'lifted' from nearby skull X and missing dimensions from skulls II and XII. And a toothless lower jaw found 25m higher in the deposit was 'attached'. From this he created a whole skull from which, in turn, a sculptress resident in Peking at that time, Mrs Lucille Swann, created a fictional 'Nellie'. Weidenreich commented on the model, "the most striking peculiarity is... the thickness of the neck." There was no evidence whatsoever for this imaginatively brutish characteristic, yet it is Nellie's profile which often illustrates *Homo erectus* in school and college textbooks as part of the anthropoid sequence supposed to lead to *Homo sapiens.*

By 1943 Weidenreich had written his extensive 'The Skull of *Sinanthropus*'. His colleague von Koenigswald commented:[65] "I believe that many people who have admired the splendid drawings and photographs in his books would be disappointed if they saw the originals".

Would the Cenozoic Laboratory's collection of fossils have withstood wider, international analysis and criticism? Sadly, we shall never know because it almost completely disappeared during or just after the war. Some say fossils were captured by Japanese on a train to the coast from Beijing while in transit to America but, on the unfortunate day of Pearl Harbor in 1941, these same troops destroyed them. Others assert that the couple or more boxes were lost in the general fog of war. Or were they held by Tokyo University at the end of the war? In a letter to the New York Times (22-3-51) Mr. Pei, now working under Communist rule, claimed they had been shipped from there to American Museum of Natural History. This was denied.

On the contrary, far from it being sent to America Rev. O'Connell believed that Pei destroyed the evidence after Weidenreich left in 1941 and before the Chinese Government returned to Peking. This was to avoid an accusation of fraud because the models did not correspond with descriptions of fossils published by Boule and Abbé Breuil. Mr. Pei did, in fact, resume his work of excavation under the Communists and in 1966 a skull, with an extremely large gap between front and rear pieces, was hailed by the Chinese authorities as that of Peking man.

O'Connell's interpretation of the facts was that several thousand years ago a large-scale industry of quarrying limestone and burning lime was carried on at Choukoutien. It was carried on at two levels. The lime was burnt by grass, straw and reeds (as it still is where coal is lacking in China), leading to great quantities of ash. Thousands of quartz stones were brought from a distance to construct lime-kilns; they were found at both levels with a layer of soot on one side. Such large-

---

[65] Meeting Prehistoric Man p.55.

scale lime-burning, it was presumed, serviced the needs of the ancient city of Cambulac on the site of the present city of Peking.

The circumstance of the Cenozoic Laboratory's lost treasure is shrouded in the mystery of rumour, innuendo and contradictory accounts. Tellingly or not, Teilhard de Chardin never gave his version of the event and the authorities still seem uninterested.

One loose thread. On 16-17 December 1929 reports appeared in the *Daily Telegraph* and *New York Times* of ten skeletons that had been found at Choukoutien. *Nature* announced the discovery on 28 December, saying that Davidson Black would make an important statement on the 29th. No more was heard from anyone on the subject. Was it a mistake? Were the skeletons found to be human? Why have the reports not been queried? Surely Teilhard or Davidson Black could have given us a clue?

In conclusion, it may be regarded as probable that many of the *Sinanthropus* bone fragments at Choukoutien were derived, as Boule and others suggested, from gibbon and macaque skulls. However, human bones, signs of extensive industry and tools were also found on site (e.g. in the so-called 'upper cave'). Were these of worker families? So should we conclude that Peking Man is, like Java Man, an example of *Homo erectus*? Milford Wolpoff calls[66] for an admission that the *erectus/ sapiens* boundary is arbitrary and *erectus* should be 'sunk' into *sapiens*. And the advice of Professor Weidenreich, who took over from Black, is that "it would not be correct to call our fossil '*Homo pekinensis*' or '*Homo erectus pekinensis*'; it would be best to call it '*Homo sapiens erectus pekinensis.*' Otherwise it would appear as a proper 'species' different from '*Homo sapiens*' which remains doubtful, to say the least."[67]

What, therefore, is the essential difference between *erectus*, *sapiens* and, as we have seen, Neanderthal? Although the evolutionary paradigm of course resists, aren't these simply variations on the single, human archetypal theme?

[66]   Wolpoff *et alii*: 'Modern *Homo Sapiens* Origins: A General Theory of Hominid Evolution Involving the Fossil Evidence from East Asia' eds. Smith and Spencer p. 465-66.
[67]   Franz Weidenreich: from 'The Skull of *Sinanthropus pekinensis*' quoted by Milford Wolpoff *et alii* as above p. 466.

18. Other erecti (upright men)

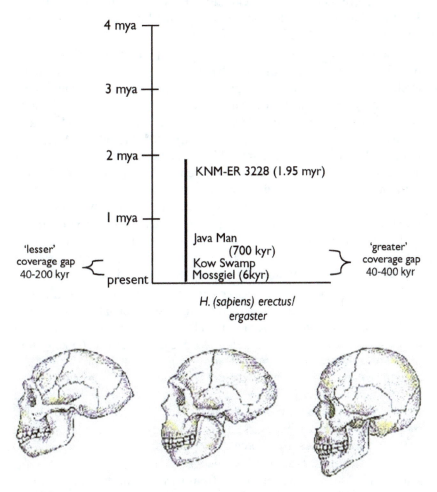

a. *H. erectus*   b. *H. neanderthalensis*   c. *H. sapiens*

18.2 You, You and You

Aren't the labels (*H. erectus, sapiens, neanderthalensis*) simply a function of evolutionary aspiration? If so, how serious are objections to this assertion? For example, is not *erectus* simply a smaller version of Neanderthal and, to all intents and purposes, within the range of human being? An evolutionary continuum is *de rigeur* but, as climatic and other factors may have forced human variations in the past, what of future pressures (not least man-made) that may vary our current races of *Homo sapiens*? Will such variations then be classified outside our own?

There is always heated, nit-picking argument among experts but it is generally agreed that over 280 fossils labelled *Homo erectus* have been found at more than 80 sites. Generally they are dated between ~400 and 1400 kyr but over 170 fossils lie outside this bundle - over 140 dated younger and over 30 older.

About 40 *erectus* sites have yielded stone tools (in fact almost every type has been found with each 'species' of human). Just as human intelligence, creative and manipulative skills vary so does the fashion and quality of such tools. They are found worldwide and some authorities have dated the oldest provenance at 3.4 myr. Hand-axes (perhaps killing missiles) are widely distributed throughout Africa and Eurasia. Tools in Indonesia and China have been track-fission dated to nearly 1 myr. Stone choppers used around Lake Turkana are identical with ones used today and 8 *erectus* sites have yielded promethean evidence of managed fire. The oldest of these is at Swartkrans near Pretoria (~700 kyr). Although signs of shelters built possibly half a million years ago have been found at Chichibu, Japan and industry at Choukoutien in China as yet, perhaps understandably, there's not much sign of boats, canoes or coracles beyond ten thousand years ago.

How does the race (or species) of *H. erectus* compare with Neanderthal? In both the skull is low, broad and elongated; they have a thick cranial wall and the same cranial morphology but adult capacity (brain size) is less at 780 - 1250 cm³ as opposed to 1200 - 1650 cm³ (and ourselves at 900 - 1800 cm³). Bear in mind that it difficult to obtain accurate volumes from fragments and best-attempt reconstructions. Both *erectus* and Neanderthal have eyebrow (supra-orbital) ridges, variable teeth sizes, weak chins and to the rear 'pointy' slight protrusions. The post-cranial (skeletal) bones are in each case heavy but otherwise the anatomy is not noticeably different from our own.

Erecti fragments 1.4 myr or younger include the Trinil/ Sangiran skulls (Java Man), a dozen or more from various Chinese sites and over 15 from Africa. At Swartkrans (near Pretoria) bones aged 1.5 to 1.9 myr were attributed to the *erectus* clan. These were found with bone tools and evidence of the use of fire. Anatomist Le Gros Clark later rejected the classification. Remember the Javan Modjokerto infant at 1.9 myr? There is also a hip-bone, **KNM-ER 3228** dated at 1.95 myr. This fossil is similar to Arago 44 (from Tautavel in France)[68] and Olduvai (OH 28) from the Leakey collection and dated at ~ 1 myr. A million years with no evolution! KNM-ER 3228, so obviously not fitting the evolutionary

---

[68]   site of *erecti* fossils dated ~450 kya; however, in 2015 an older human (*sapiens*) tooth dated ~550 kya was found there. Nature (14-10-15) also reported 47 human teeth from Daoxian, China dated ~100, 000 kya.

continuum, was therefore reclassified into a doubtful taxon to be dealt with later and called *Homo habilis*.

At the low end more than 70 *erectus* fossils are dated more recently than 30 kyr with the youngest at only 6 kyr - dates that correlate with Garniss Curtis dates of Javan fossils from Ngandong (27-53 kyr). An Australian specimen, the Cohuna cranium which some have considered aboriginal, has Ngandong morphology. Indeed, Franz Weidenreich commented that *Homo soloensis* (Java Man) traits are found in quite modern aboriginal skulls.

Other so-called robust but human-like 'down-under' fossils have been gathered, mostly since the 1960's, from Willandra Lake (30 kyr), Talgai (12 kyr), Coobol Crossing and Kow Swamp (~9 kyr). Two relatively complete crania from Kow Swamp (KS1 and KS5) were classified *H. erectus*.[69] To these add the Mossgiel cranium and (from Western Australia) Cossack skull both dated ~ 6 kyr. *Of course, this doesn't fit the story.* **If KNM-ER 3228 *is* H. erectus and the species has lasted unchanged for 2 myr then where, long time passing, has all the evolution gone? Moreover, the idea of co-existent *sapiens* and *erectus* populations severely offends a time-and-species-splitting evolutionary model.** Happily, for what it is worth, for now an Out-of-Africa scenario dictates that *H. erectus* never reached Australia and so the remains *must* be reclassified as aboriginal *H. sapiens*.

Has *DNA* analysis helped to resolve the issue? In some *erectus* fossils the mt-*DNA* is similar to human. Does this mean *erectus* and *sapiens* Australians mixed (as did *sapiens* and Neanderthal far to the west) to generate the genetically modern race of aborigines? Is today's human simply a viable hybrid or a part of on-going variation on the archetypal human theme?

One discovery, **Mungo Man 3** from Mungo Lake in the Willandra region, is anatomically modern but has dissimilar mt-*DNA*. What *is* the range of mt-*DNA* variation possible within *H. sapiens* (which perhaps includes the *erectus* and Neanderthal labels)?

H. erectus diagnostics include generally thick bone, human size, shortish arm-bones, large teeth, chinless jaw, a receding forehead and brow ridges bulgy but less so than Neanderthals. These features are present in varying degree with the Australian specimens so far discovered. Perhaps, in these young fossils, the thick skulls were due to nutritional problems, local survival of the fittest or disease. But if human skulls can vary in this way why (except for evolutionary purposes) invoke the classification *erectus* in the first place?

---

[69]   Nature Vol. 238, 11-8-72 ps. 316-9.

Do humans have sloping, receding foreheads? The typically 'primitive' recession is justified (in the Australian case but not elsewhere in the world where *erectus* fossils are found at 'correct' ages) by using the contrivance of 'artificial cranial deformation'. This aberrant behaviour is not part of modern aboriginal culture but for some unexplained reason it is explained that some of their Kow Swamp forebears were, as infants, bound long-term or repeatedly pressured by motherly hands in order to have their thick skulls deformed. The consequent recession *appears* pre-human but, since the pates are actually human, theory's honour is preserved. Except that if morphology alone can show what evolution should why is the latter needed? Does such plausible interpretation help explain or cleverly explain away the facts?

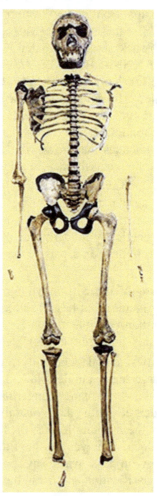

18.3   Turkana Boy (KNM-WT 15000)

Ergaster? In Africa the correlate of *H. erectus* is called *H. ergaster* and assigned a time-line of ~1.8 to 1.3 mya. *Ergaster* ('working man') sports greater facial height, less pronounced brow ridges and a taller cranial vault with thinner bone. (S)he is variously, depending on who you talk to, considered a variant or the direct ancestor of *H. erectus*. Examples include a number of fragments (1.6-7 myr) from Olduvai, **KNM-ER 3733** (~1.7 myr, a supposed female from Koobi Fora on the eastern bank of Lake Turkana, formerly Lake Rudolph) and twelve year old '**Turkana Boy**' (~1.6 myr, also called 'Nariokotome Boy' and **KNM-WT 15000**) found in 1984 west of Lake Turkana. The woman's skull, of tall cranium, is smaller than a female pygmy (~ 850 cm³); today *H. sapiens* pygmy populations are distributed across Africa, South America, S.E. Asia and Australasia. The boy's skull volume is estimated at ~920 cm³ but the rest of his skeleton is quite complete and indicates (where *erectus* is supposed to be a pygmy-sized 'human') an adult height of over six feet. Turkana boy's anatomy is like *H. sapiens* but, given that a fertility test is impossible we cannot, in this most basic sense, tell the *erectus*/ *sapiens*/ *neanderthalensis* grouping as divisions apart or, set outside the evolutionary mandate, as a union composed of genetic variations of ourselves.

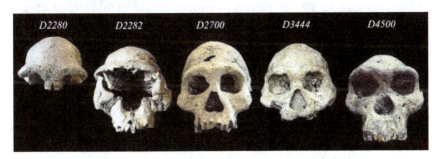

18.4   The Dmanisi five.

If they are found in different places skull variations attract classification into different species. Found in the same level and location they need necessarily be identified as variations of the same species. The latter applies to five most interesting skulls[70] unearthed between 1991 and 2005 with accompanying bone fragments and crude stone tools at **Dmanisi** in Georgia CIS. The bones include an ape-like femur, a tibia, kneecap, ankle-bones, collar-bones, upper arm bones, vertebrae and part of a shoulder blade. The range of skull cranial capacities from ~550 cm³ (D4500) to 780 (D2280) is as large as that

---

[70]   Science Vol. 342 (18-10-2013 ps. 326-331).

estimated for African *H. erectus*, *habilis* and *rudolfensis*. In some respects they resemble so-called Olduvan habilines such as OH 24, **KNM-ER 1813**, KNM-ER 62000 (the well-preserved face of a juvenile) and, found in 1972, **KNM-ER 1470 (*H. rudolfensis*)**. The latter was theoretically but not physically associated with a modern femur and lower leg fragments dug from the same site and level. At 2.6 myr KNM-ER 1470 was at first labelled human by Louis Leakey. This anomaly was soon revised to ~1.9 myr and species *H. habilis*. It is now called *H. rudolfensis* (see next Chapter).

Some Dmanisi skulls appear simian and others more human. D4500, associated with a large mandible (D2600), has a large face and complete cranium (~ 550 cm³) and looks simian. Three range between 600 and 650 cm³. Of these one (D2282) looks simian and two ((D2700 and D3444) more human. Likewise the high cranium with no face or jaw (D2282 and at ~780 cm³ of *erectus* size). Same level, same location but large variation - what are we to make of this?

Is it age, sex or what? Such extent of variation is found in both chimps and modern humans. Some claim the Dmanisi five range across so-called *habilis*, *erectus* and *rudolfensis* species put together! In the genus *Homo* cranial capacity (which differs slightly from brain size) runs from ~780 cm³ (*erectus*) through to 1650 cm³ (Neanderthal). *Sapiens* variation runs (excluding the pathology of, say, microcephalic idiots) from about 900 to 1600 and even larger. This a range of over 800 cm³ which might be matched by the others if the same number of specimens were available.

So are the Dmanisi gang a collection that represents an isolated population of humans (a sub-species of *ergaster* called *H. georgicus*[71])? Or, dated the same as 'archaic human' *H. rudolfensis* (~1.8 myr), did they issue as a first and very, very early wave of humans 'Out of Africa'? If so, did the Javan *erecti* (at ~0.7 myr dated over a million years later) 'devolve' or 'reverse evolve' into a more 'primitive' form? In other words, do the Dmanisi five represent ancestors of Asian *erecti*, Neanderthals and thereby *sapiens* (ourselves)? Of course, hybrid viability cannot be tested but, if it happened, might not *erectus* genes be found in you and me? Or is this species simply an extinct one?

---

[71] also *Homo erectus ergaster georgicus*, a conflation which demonstrates certain confusion.

19. Broken Hill Man

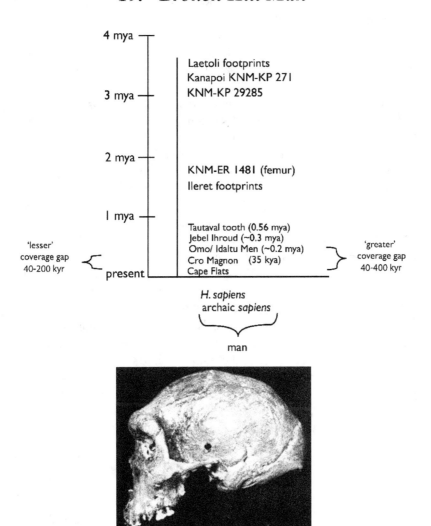

4 mya

Laetoli footprints
Kanapoi KNM-KP 271
3 mya KNM-KP 29285

2 mya

KNM-ER 1481 (femur)
Ileret footprints

1 mya

Tautaval tooth (0.56 mya)
Jebel Ihroud (~0.3 mya)
'lesser'   Omo/ Idaltu Men (~0.2 mya)   'greater'
coverage gap   Cro Magnon   (35 kya)   coverage gap
40-200 kyr   Cape Flats   40-400 kyr
present

H. sapiens
archaic sapiens

man

19.2   Broken Hill Man (*Homo rhodesiensis*)
with so-called 'bullet-hole'.

Do philosophical and/ or political principles ever triumph over scientific ones? **Broken Hill Man** (*aka* Kabwe Man, Rhodesian Man or ***Homo rhodesiensis***) was found in a South African copper or heavy metal mine and reported in Nature (Nov. 1921) by Dr. Arthur Smith Woodward FRS, a senior palaeontologist at the British Museum (Natural History) and actor in the unfortunate Piltdown theatre. The skull (~1280 cm³) has large brow ridges. Surely this 'savage' mien could not cry out Neanderthal in Africa? Stone and bone tools, bolas (used today for catching cattle) and various animal bones (including lion, leopard, elephant and extinct long-horned buffalo) were found but inconsiderate mining personnel, in the course of excavations, destroyed the site. The skull, of which Elliott Grafton Smith made an endocast, was not fossilised nor even, surprisingly, mineralised. Smith Woodward put the age of an 'archaic man' at ~10 kyr.

You and I can't age that fast but Broken Hill Man was accelerated in a time capsule! By 1965 Richard Klein had, giving various reasons, lifted Carlton Coon's new date[72] of 40 kya to a minimum of 125 kya. Then, in 1999, American palaeontologist Ian Tattersall lifted its antiquity to 300 or 400 kya. This was without much reason except perhaps a perceived correlation with nearby **Saldanha Man** (a skull variously dated between 100 and 500 kyr); or to facilitate the 'Out of Africa' notion wherein modern men are supposed to have emigrated from that continent around 150-200 kya and taken over the world. The classification of extinct hominin (*Homo* and his antecedents) has hung on and this, to date, is as far as his time flies. It's still very rapid aging; around 350 kyr in little more than 0.08 kyr (80 years) without good evidence he lived more than perhaps 1 kya! It's called 'philosophical boost'.

Actually, quite a number of fossils of so-called '**archaic men**' have been found with anomalous (*aka* 'theoretically inconvenient') dates. The 'classification' amounts to a basket into which to throw elderly fossils of possibly *erectoid*, *Neanderthal* or *Sapiens* cast. Such 'transitional' skulls generally have your own cranial capacity ((1100-1300 cm³), beetling brow ridges, low front-sloping and lack of Neanderthal protrusion (bun) at the rear. Their post-cranial bones appear human or not depending on who describes what.

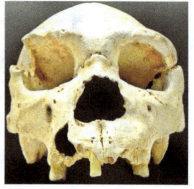

19.3 *H. heidelbergensis* (Mauer jaw and Sima de los Huesos skull).

---

[72]   see Chapter 24: Carlton Coon's model of human evolution.

In this respect is not, as some moot, *H. rhodesiensis* the same as (or a sub-species of) **H. heidelbergensis**, the Heidelberg Man found in 1907 near that city in Germany. This taxon would then, between perhaps 350 and 150 kya, have roamed Africa and, depending on when you date the presumed exodus from Africa, Europe and Western Asia. It is also believed by some that Neanderthals and humans evolved, say, 130 kya, from this sort of hominin. (S)he averaged human height (taller than Neanderthals at about 5 feet 9 inches and had a brain (1100 to 1400 cm³) your size.

There have been, as well as *H. heidelbergensis*, a number of *sapiens*-like remains found 'out of time'. This inevitably leads, under one pretext or another, to their dismissal, temporal 'rearrangement' or assignment to a 'suspense account' called '*Archanthropus*' or '**archaic *Homo sapiens***'. Examples include, among many others (see Index), 'Swanscombe', 'Boxgrove', 'Fontechevade' and 'Vertesszöllos' Men. Swanscombe Man (actually a female) dated at ~400 kyr consists of a couple of skull fragments in the company of numerous stone and flint tools (including famed Acheulian handaxes). At Boxgrove in Sussex the tibia of a human dated 500 kyr and similar tools came to light. Close examination of them shows *H. heidelbergensis* 'Boxgrove Man', said by some to be the oldest person in Britain, was at least as intelligent and skilled as any modern knapper (*H. sapiens*)! Classification is not simple because experts can disagree. For example, around 1970 a decision was made to withdraw the *H. erectus* label wherever stuck on European finds (like some of those above) because the fellow was not supposed to have reached here yet. Candidates were re-allocated to Neanderthal or 'archaic human' types.

We can reasonably ignore as Miocene a graveyard found in 1812 in Guadeloupe. However, claims have been made for human finds in the Pliocene era. These include skulls found at Castenedolo in Italy and Calaveras in California. In the first case, at **Castenedolo** near Brescia in Italy, Professor Guiseppe Ragazzoni, a geologist, found a modern skeleton in undisturbed blue clay also dated Pliocene (over 5 mya). The find was confirmed by anatomist Guiseppe Sergi. Careful inspection of the overlying rocks showed them undisturbed and anthropologist Sir Arthur Keith admitted: "As the student of pre- historic man reads and studies the records of the Castenedolo finds, a feeling of incredulity rises within him. He cannot reject this discovery as false without doing injury to his sense of truth, and he cannot accept it as fact without shattering his accepted beliefs."[73] However, 'injury' was preferred to 'shattering' and Ragazzoni's find was dismissed as 'bad archaeology'.

Never fear because, according to Boule, an official report on these skeletons in 1899 noted that all the fossils from the deposit except the human

---

[73]   Keith A. *The Antiquity of Man,* Williams and Norgate, 1925, p. 334.

ones were impregnated with salt. This might imply that they are from relatively recent burials. And the discovery of a second skeleton in a nearby fissure, this time clean of encrustation by marine material, led directly to a notion that all the bones were from a medieval cemetery; later carbon-14 dating of ribs (of the second skeleton?) precisely confirmed this theory. The acceptance of a Pliocene date for the Castenedolo skeletons would have created so many insoluble problems that one might hardly hesitate in choosing between the alternatives of adopting or rejecting their authenticity. Who, after all, would wish 'bad' archaeology might nip a theory in its fragile bud? For sure, intrusive burial! Castenedolo case rejected!

At **Calaveras** a gold-miner, digging in 1866, unearthed a human skull thickly encased in cemented gravel from 130 feet below the surface. The state geologist, Dr. Whitney, believed them genuine; another professional geologist, Clarence King, gouged a stone grinding instrument from local rock dated at 9 myr; and more artefacts and fossils were discovered in the area. Anthropologist William Holmes from the Smithsonian Institution soon weighed in. Alfred Russell Wallace, co-formulator of evolution theory, expressed dismay at the way anomalous evidence for modern man in Pliocene rocks was 'attacked with all the weapons of doubt, accusation and ridicule'. Nevertheless Holmes castigated Whitney for his lack of caution in presenting evidence that contradicted theory; anecdotal evidence whipped up a claim that the illiterate but Darwin-savvy miners had, by moving the skull from a local cave and cleverly embedding it the mine, been persuaded to commit this funny hoax for free! Later radiocarbon dating was said to have dated the skull at only a thousand years; and other finds in the area, found some fault with, were equally and summarily dismissed as plants, hoaxes or intrusive burials.

And the same went for the clay-filled **Olmo** skull found 1883 at a depth of 15 metres in a railway cutting. The date was reduced to ~75 kya.

No doubt, in the cases of Java, Piltdown and Peking men, anthropologists were not as critical if the evidence were in favour of Darwinian hypothesis - far from it. Blatant fraud went unchecked for decade after decade; skulduggery and manipulation with respect to reconstructions, inconvenient dates and taxonomic classifications abounded; much propaganda value, especially illustrative, was extracted from scandalously ill-defined material; the complaint of Wallace was true; and such monkey business may, in the highly emotional field of human origins, still occur.

The human **Sondé tooth** (~700 kyr) was found by the Selenka-Trinil expedition at the same level as Java Man's human femur. Frau Selenka therefore concluded that Java and *sapiens* were contemporary. But her reports were almost entirely ignored by the scientific community as 'incorrect'; such 'incorrectness' has been justified by a claim that the tooth was large and therefore actually *erectoid*. Despite this, in 2015 a human incisor (Arago 149 dated ~550 kyr) was unearthed at Tautavel in Southern France.

At **Vertesszöllos**, Hungary in 1965 a few tooth fragments and part of an adult cranium were lifted and dated ~300 kyr. The latter fragment is very thick and broad, with a mixture of modern and primitive features. This is considered characteristically Neanderthal/ archaic *sapiens*. Its age, which has variously been estimated to be from 160 to over 350 kyr, would match such assessment. Similarly tools and two *sapiens*-like partial skulls lacking brow ridges (~160 kyr) were found in 1947 by experienced excavators led by Germaine Henri-Martin at the Fontéchevard cave in France. The age of the deposit (~100 kyr) was defined by the animal types it contained; because it had been sealed by a thick, undisturbed layer of stalagmites, later burial was impossible. Vallois reckoned they were of Swanscombe type (i.e. Broken Hill and Heidelberg) and of the same biological population - although the second skull was thicker. Other experts argue with his conclusion. What are we to believe?

19.4   Petralona skull transfixed by stalagmite.

In Greece we find, accompanied by tools including bone awl and evidence of controlled use of fire, the **Petralona skull** (*Archanthropus europaeus petraloniensis* to the experts). An age estimate of ~70 kyr was later measured, by electron spin resonance, at ~200 kyr or (as claimed by its finder, Aris Poulianos, using the same method) ~700 kyr. A couple of skeletons found later in the cave racked up 800 kyr. The skull was originally called *H. erectus* but a decision by Clark Howell that *erectus* never existed in Europe meant it was reclassified (with Vertesszöllos and co.) as archaic *H. sapiens*. When, however, Out of Africa theory meant that European humans were not related to but killed off by migrants from Africa, then 'archaic *sapiens*' was turned into *H. neanderthalensis*. There Petralona, swept by the flow of theory, rests for now. Much the same analysis applies to Italian Altamura Man (found 1993, dated possibly ~250 kyr) whose body was probably literally washed by the sweep of rainwater into his well-preserved skeleton's limestone-encrusting mortuary chamber.

Stenheim Man (below) is also dated from 250-300 kya. Its cranial capacity is ~1150 cm³ (human, though some estimate sub-human at 950 cm³) and its classification *H. heidelbergensis*. *H. heidelbergensis* is considered by some to be pre-Neanderthal in the line towards *sapiens*.

19.5   Stenheim skull (side and front).

Is there any more time for archaic humans, other than Broken Hill Man, in Africa? There is. A complete skeleton was found by Hans Reck at Olduvai Gorge in 1913. The first Olduvan fossil, it was laboriously chipped from Bed II's hard rock confirmed at ~0.5myr - far too old for evolutionary *sapiens*. Both Reck and Louis Leakey[74] state clearly there was no evidence of intrusive burial though Leakey doubted such age. At a later date he and colleagues, including Sir Vivian Fuchs and Arthur Hopwood from the Natural History Museum, revisited the site and confirmed the 'non-burial' assessment[75] but re-dated to ~700 kyr (as Java Man). The anomaly then, in 1932, created a flurry of correspondence in Nature. During this time Professor Percy Boswell received samples of material brought back by Louis Leakey from several layers on the site. Subsequent analysis by Berlin Professor Mollinson seems to have employed a mixed bag as he reports elements from Beds V, IV and III above as well as the fossil's own Bed II below. Whether misleading or not these tests convinced Reck and the others to recant and adopt the 'intrusive burial' refrain. By 1974 the skeleton had disappeared (!) but, using the skull, the accuracy of Mollinson's tests was criticised. Radiocarbon measurement demonstrated an age of ~20 kyr or less although, admittedly, inaccuracy or contamination may have occurred due to using samples of calcrete from Bed V, freshwater shells and only a small amount of fossil bone. This bone was found at the boundary of Beds

---

[74]   Nature Vol. 121 (31-3-28 p. 499).and Nature Vol. 131 (1-4-33 p. 477).

[75]   Nature Vol. 128 (24-10-31 p. 724).

III and II and thus laid, it was presumed, 400 kya. Clearly, *H. sapiens* of 400 kyr is unacceptable from an evolutionary point of view but it has been claimed that older deposits were exposed by faulting (Johanson and Shreeve 1989) so that intrusive burial could, despite the early assurances, have occurred; also (by Morell, 1995) that Bed III is of recent origin. Thus Reck's very early human has lately been, it may appear, comprehensively dissed!

The Leakey family has, supported by the National Geographic Magazine, famously collected fossils from Kenya and Tanzania. A Kanam Jaw and Kanjera skull fragments (contemporaries of Swanscombe Man) were discovered by Louis Leakey near Lake Victoria in 1932 and claimed by him to be a very old but anatomically modern human ancestor viz. modern humans buried in older sediments. Unfortunately Leakey failed to document the exact site. His confused evidence meant that a 27-man committee was highly critical of the archaeology.[76]

Most people learn from their mistakes and the Leakeys went on to produce a mass of evolution-interpreted material from Olduvai. This includes some old material and, you may have begun to observe, with more interested parties than fossil fragments populating the subject, how speculative an art is palaeoanthropology. There is little consensus as to family trees but, when it comes pieces that, by scientific date and character, could threaten evolutionary paradigms you might expect them to attract fierce materialistic, pseudoscientific exertion to revision in order that 'creationists be silenced'! This, translated, means that all philosophy except materialism must be absolutely dissed!

Old (therefore perhaps anomalous and debateable) material includes fossils from **Moroccan Jebel Ihroud (at ~ 0.3 mya), the Omo and Herto Ethiopian material (~0.23 - 0.16 mya), KNM-ER 813, 1590, 1472, 1481 (~1.9 myr, femur and leg bones), 1470 (~1.9 myr, cranium), KNM-KP 271 (~3.5 myr, upper arm bone fragments), KNM-KP 29285 (~3.5 myr), Modjokerto child and human footprints from Koobi Fora (Ileret) and Laetoli (~1.6 and 3.6 myr respectively)**. And if **Omo Man** (~ 195 kya) is, in concurrence with Out of Africa theory, identified by some as the oldest AMHS (anatomically modern *Homo sapiens*) you will already spot the jarring anomaly.

Found in 1932 near Bloemfontein, by 2000 the slightly brow-ridged **Florisbad skull** fragment had been raised from an estimated ~40 to ~260 kyr. Similar in reconstructed appearance to Stenheim Man (*fig. 19.5*), this skull is, with expert confusion, labelled *H. helmei*, *H. sapiens* or *H. heidelbergensis*! Next, between 1967 and the mid-seventies near the Omo River in S. W. of Addis Ababa in Ethiopia, two partial skulls, a few jaws

---

[76] Nature Vol. 135 (9-3-35 p. 131): 'Human Remains from Kanam and Kanjera' by Percy Boswell (see also Reck's Olduvai Man).

and a couple of hundred teeth were dug out by Richard Leakey. In 2008 a tibia, fibula and talus (~0.2 myr) were also uncovered; these leg bones are like those that compose **KNM-ER 813** (~1.8 myr, foot and leg bones). And modern-looking Omo 1 (0.23 myr) is perhaps the oldest accepted specimen of *H. sapiens* (his 'brother', Omo 2, sports an 'archaic' appearance reminiscent of *H. heidelbergensis* or Java Man).

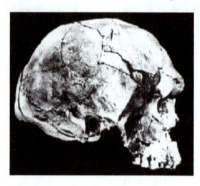

19.6   Idaltu Man (see also Chapter 24: 'photofit').

At Middle Awash N.E. of Addis Ababa in 1997 Tim White found three crania but no post-cranial parts - two adult males (1450 cm³) and a young child called **Herto or Idaltu Man** (*Homo sapiens idaltu*). At ~0.16 myr these shiny-pated chaps were younger than their Omo neighbours but of 'archaic' countenance. Stone tools were scattered round. Was Broken Hill Man father of them all?

Omo and Herto candidates are placed and dated well in line to make fine candidates for sally Out of Africa. They represent manna in a fossil desert on the Exodus from Africa so that its region of N.E. Africa has been called by Mitochondrial Eve theorists 'The Cradle of Mankind'. But nothing too much older, please, than these two Ethiopian types.

19.7   KNM-ER 3733

Human approximates **KNM-ER 1590 (~1.85 myr, partial cranium and almost complete dentition), 1472 (~1.9 myr, femur) and 1481 (~1.9 myr, femur and leg bones), 3228 (~1.95 myr, hipbone, see previous chapter) and 3733 (~1.7 myr, cranium about 850 cm³, see also previous chapter)** have each been assigned, by one expert or another, more than one classification but have, given their age, wound up as *H. erectus* or *H. habilis*.

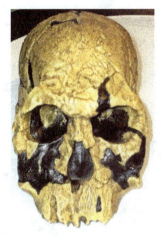

19.8   KNM-ER 1470 (*H. rudolfensis*)

KNM-ER 1470 (see also previous chapter) is in another league. The skull, reconstructed from many smashed fragments, was discovered in 1972 at Koobi Fora by a Leakey dig financed by the highly evolutionary-minded National Geographic Magazine. At first classified *H. habilis* although it was of larger cranial capacity (in *erectus* range of ~800 cm³ according to the possibly incorrect but first, media-fanfared reconstruction), with a tall, rounded cranium, virtually no brow ridges and found with human leg bones of similar age nearby.[77] Roger Lewin noted the uncertainty of angle of attachment of the upper jaw to face; according to preconception you could hold it forward to make a long face or back so that appeared short. You could make it ape or man-like!

It was originally dated at nearly 3 mya, a figure that caused much confusion as at the time it was older than any known australopithecines, from whom *habilis* had supposedly descended. A lively debate over the dating of 1470 ensued (Lewin, 1987; Johanson and Edey, 1981; Lubenow, 1992). It has fallen 40%, for now, to a 'more reasonable' ~1.8 mya.

The braincase seems surprisingly modern in many respects, much less robust than any australopithecine skull, and also without the large brow

---

[77]   Nat. Geog. Magazine Vol. 143, no.6 (June 1973) ps. 819-29.

ridges typical of *Homo erectus*. The face, in contrast, is large, robust but (as *H. sapiens*) flat with receding orbital bones. Called *Pithecanthropus rudolfensis* until 1986 when it was re-christened (in an evolutionary sense) *H. rudolfensis. If* 1470 is related to the distorted fossil skull KNM-WT 40000 (*Kenyanthropus platyops*) to which it has some claimed resemblances, it may eventually be reassigned to the genus *Kenyanthropus*. However, the morphological interpretation of early hominids has swung considerably over the decades.[78] And further reconstructions up to 1992 have (excepting brain-size) swung towards an antique australopithecine representation (more like KNM-ER 1813 and 3733).

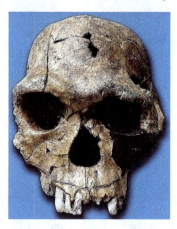

19.9   *Homo habilis, erectus* or 'archaic' *sapiens*?
Trans-special KNM-ER 1813 est. 1.8 myr
(see also Chapter 'Muddle in the Middle' *fig.* 21.4).

Thus evolutionists and non-evolutionists both seem divided on whether **KNM-ER 1470, KNM-ER 1813** (~1.8 myr, found in the same place next year and with similar cranial morphology except smaller capacity and more pronounced brow ridges) or **KNM-ER 3733** are simian or human. But could not the capacity for, say, brow ridges exist within the *DNA* of *H. sapiens* adaptive potential? So should 1470 not be called *Homo*? And thus have lived, labelled (where brain size does not equate with intelligence) *erectus* or *sapiens*, as capable a human life as we do? Or should such skulls be removed to the simian end of 'Muddle in the Middle' (Chapter 21)?

Back to the problem of succession - how could humans be that old? Or older? At 4.5 myr the lower (distal) end of an upper arm, **KNM-KP 271** from Kanapoi in Kenya, is both offensive, contentious and

---

[78]   e.g. T. Bromage 'Faces from the Past': New Scientist Vol 133 (1803) 11-1-92 ps 32-5.

unaffordable. Found in 1967 the elbow most digs whose ribs? At first it was diagnosed using multivariate analysis with chimp and australopithecine upper humerus and found to be 'strikingly close to the means of the human sample'. Another study found it 'indistinguishable' from modern *H. sapiens*.[79] Yet it was promptly despatched to the species of *Australopithecus africanus* (and later *A. anamensis*) because no other hominids are known from 4 million years ago!

Besides the 'wrong' time element it is argued that this decision is plausible because, as proximal (upper) end of an australopithecine humerus is visually similar to a human's, you could guess the base (distal) end must be too. Or, because the lower humerus is by itself a poor diagnostic indicator, it would be premature to claim that KP 271 could not be australopithecine. Or perhaps the advanced statistical technique multivariate analysis is unable to completely distinguish between human and chimp populations. Thus, in 1996, a further and extensive study of the lower humeri of apes, humans, and hominid fossils re-using multivariate analysis was performed by researchers Lague and Jungers. This time round the reverse was demonstrated viz. that KP-271 lies outside the range of human specimens. Instead, it clusters with a group of other hominid fossils so strongly that the probability that it belongs to the human sample, rather than fossil hominid group, is less than one thousandth (0.001). Thus the new classification *A. anamensis* (by Maeve Leakey) was justified. Is the label *Homo* dissed for good or is it time for yet another analytic dig?

All this intellectual heat while prehistoric man is skipping! The Ileret (Koobi Fora) footprints (~1.56 myr) show walking and running in our style although they are carefully, compulsorily ascribed to extinct hominins such as *H. ergaster erectus*.

Previously, after Dr. Louis Leakey's death in 1972, his wife Mary continued excavating both at Olduvai Gorge and further south at Laetoli. In 1976 she came across trails (~3.6 myr) set in Laetoli tuff. She also discovered, amid many bird and animal tracks identical to modern types, a trail of ape-like footprints in volcanic ash considered to date from nearly four million years ago. The big toes are not simian, i.e. opposable. From slides and photographs these tracks appear pretty much as yours or mine would if we walked barefoot across such a surface. They are 'free-striding' and indicate an upright gait with one foot crossing in front of the other in the manner of a hunter. They could easily have belonged to a modern girl or a pygmy. The shortish stride could have arisen because the person was walking cautiously across damp volcanic ash. Everyone who examines them agrees they are

---

[79] Science 190 (31-10-75 p. 428).

sapiens-like in stride. Even a digital expert, evolutionist Russell Tuttle, honestly reflects Mary Leakey's initial assessment that the latter footprints are remarkably similar to those of modern man.

Nevertheless, theory commands that commentary employs the term 'hominid' rather than 'human'; and that the prints are interpreted as the tracks of a small-brained, 'upright', three-foot ape whose bones were found nearby - *Australopithecus afarensis*! However, believing they are too human-like to belong to *A. afarensis*, Tuttle disagrees. He suggests they might belong to another species of australopithecine or an early (non-modern) species of *Homo*. The date keeps snagging. After extensive study he admits (Pitted Pattern p. 64) that if they 'were not known to be so old, we would readily conclude that they were made by a member of our genus, *Homo*.' But conformity with theory submits assignment of the footprints to an extinct form of ape.

Thus Don Johanson, whose 'Lucy' we shall soon meet, can think that she or her mates were skipping over Laetoli. He claims his bones were fully adapted to a modern style of bipedality and that the *A. afarensis* foot bones found at Hadar, when scaled to an individual of Lucy's size, fit the prints perfectly. Ronald Clarke and Phillip Tobias also believe that the Laetoli tracks could have been made by feet with bones similar to those of a Lucy-like organism called 'Little Foot' (Stw 753 announced in 1995 and dated by various researchers between 2.2 and 4 mya). 'Little Foot' is an Australopithecine (*A. prometheus*) that foraged in trees. Clarke matched up its fossil with bones found in museum boxes, bags and a cave at Sterkfontein. Some disagree with his assessment. Others (e.g. Stern and Susman), have argued that Lucy's foot and locomotion were bipedal but not yet fully human-like and weigh in with a belief that they know how Australopithecine footprints show subtle differences from human ones and those at Laetoli must have been made by *afarensis*. Poor running man - erased in thought if not the earth's reality because he doesn't fit a scientific scale!

Do you remember the different-looking Dmanisi crew (previous chapter) and Spanish Sima de los Huesos cave? The latter's natural crypt is, like the 'Rising Star Cave of 'Naledi Man' in South Africa, deep, narrow and with no end visible from the entrance. It functioned (~400-550 kya) solely as an ossuary in which its population, one of great morphological diversity, 'piled' their dead.

The 'archaic' bones had been disturbed by casual fossil hunters but in 1992 Juan Luis Arsuaga identified and analysed the remains of at least 33 individuals, with 3 finely preserved fossil skulls whose variation is such that, let alone the other bones, it encompasses the whole range of 'archaic' *H. sapiens* found in Europe. **Indeed, as regards 15 cranial characteristics Arsuaga's fossils show 7** similarities with ourselves, 10 with Neanderthals and 7 with *H. erectus* (supposedly the precedent of Neanderthals) - suggesting that differences may be due to non-evolutionary causes occurring in a single, unique species, *H. sapiens*.

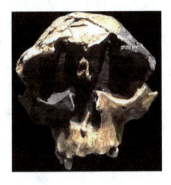 

19.10   The Boy from Gran Dolina

Just round the corner in a cave called Gran Dolina about 25 bones from various bodies can be arranged as a composite (but very incomplete) skeleton. They include a skull (cranial capacity c. 1000 cm³) of a lad interpreted by some as, at ~ 800 kyr, the first human - *Homo antecessor*. Was the Sierra de Atapuerco district one of travellers, migrants in a mixed society or folk who dropped in over an extended period of time? Last to first, first to last, all together in the after-life.

'Archaic *Homo sapiens*' is a heading under which to list fossils of humanity, some as recent as 5 kyr (Cape Flats skull, South Africa) and 140 kyr (skull fragments and jawbone of tool-producing Israeli Nesher Ramla) or others as ancient as Broken Hill Man (~400 kyr), Neanderthal, bones from Sima de los Huesos, Apidima crania, *H. antecessor* and perhaps Java Man (~800 kyr). They involve a selection of cranial morphologies, human cranial capacities (1100-1300 cm³ with exceptions both ways) and brow ridges. The post-cranial bones are essentially human (the genus *Homo* may include *sapiens*, *erectus* and Neanderthal variations).

The point is that bodies from Sierra Atapuerco, not least those bundled in the Sima de los Huesos grave, hail from a single population of men and women. If such diversity can inhabit one group at a single time what does this say for distinctions imposed due to an essentially racist, progressive theory? Line up in a crowded playground. You look 'thick' so join this line; you look 'brighter', join it further towards the front; your skull slopes and his is rounded well so we'll arrange you here and there! **Actual humanity, whose adaptive**

potential resides in the human archetype, has been divided falsely and paraded in a way imperative to evolutionary theory alone. And if this theory is 'sacred' since it's permeated intellectual bloodstream what's the antidote? How could such a powerful, universal spell be broken?

In fact, genetic isolation and adaptive potential (*SAS* Chapter 23) can account for much more precise and rapid typical variation than irrational mutation. *What, therefore, if the evolution scenario is a large-scale, humanly-devised illusion?* **What, when the veil drops, do you see?**

Although no African counterpart to the Sima de los Huesos cache has yet been dusted down, you could argue from their cranial characteristics that they should be assigned to fully human type. Do you interpret such variety in evolutionary or non-evolutionary ways? Israeli caves, Dmanisi skulls and the Spanish ossuary each indicate that variation's range might fall within our special flexibility - *except that bones can't interbreed to prove it finally.*

In the last analysis, given the background, information-rich system worked through in *SAS* and the weakness of speculative objections, why should anyone be effectively badgered by a materialistic belief system out of believing that KP-271 and the Laetoli footprints are other than what they appear - modern *sapiens* 'anomalies'; or that the Sima de los Huesos ossuary was inhuman; or that human beings were not human from their archetypal start?

20. *Australopithecines*

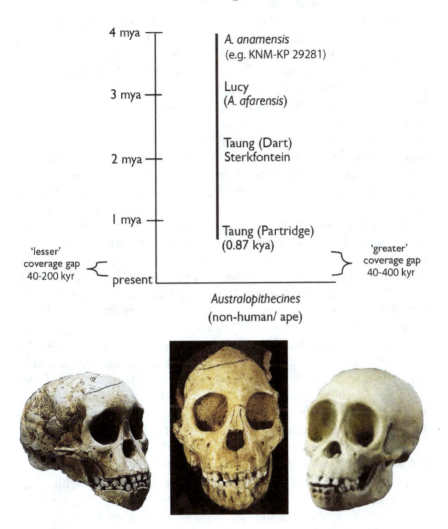

4 mya ⊤⊤

A. *anamensis*
(e.g. KNM-KP 29281)

Lucy
(*A. afarensis*)

3 mya ⊤

Taung (Dart)
Sterkfontein

2 mya ⊤

1 mya ⊤

Taung (Partridge)
(0.87 kya)

'lesser'
coverage gap
40-200 kyr

present

'greater'
coverage gap
40-400 kyr

Australopithecines
(non-human/ ape)

20.2   Taung Skull (Dart's Baby) x 2; to right, baby chimp.

From man to ape - Dart's baby. In 1924 at The Place of the Lion (Taung) in South Africa Raymond Dart, a pupil of Grafton Elliott Smith, discovered a juvenile skull with simian features but, he claimed, man-like teeth. Although most experts of the time thought it was a juvenile chimpanzee he classified it ***Australopithecus africanus*** (southern ape of Africa) with a date of 2-3 mya. A medical man, Robert Broom, agreed and started hunting fossil apes at Sterkfontein. In 1936 he found the second *australopithecine* fossil (*Plesianthropus transvaalensis* renamed *A. africanus*) which consisted of parts of face, upper jaw and braincase. Dart joined his friend to excavate and by 1938 they had found a 'robust' specimen that Broom classified *Paranthropus robustus*.

However, later on in 1973, a geologist, T. C. Partridge, showed that the Ta-ung cave from which 'Dart's baby' came could not have been formed before ~ 0.87 mya, meaning that the skull could only be that age at maximum. Such discrepant anomaly (humans were, it is supposed) on the scene by this time) provoked immediate reclassification to *habiline* or, later, *Australopithecus robustus*. Nevertheless, according to Dart's successor, Phillip Tobias, by 1973 the skull had still (despite fifty years of propaganda value) never been fully analysed or described. And surely, argued Richard Klein, a date of 2 myr is 'more reasonable' than Partridge's embarrassing chronology? Thus the latter's work has been ignored.

Let's dig deeper. Whence Dart's baby? If you believe materialistically (not necessarily the same as realistically) that men and chimps have issued from a common line then where did this, the seed that bred the child, emerge? Remember, evolution must preclude ideas. Natural forces build from simple that progresses randomly to complex, mechanistic bio-form. Darwin wrote *Descent of Man* and Bronowski *The Ascent of Man*. But, up or down, right or wrong, their magisterial theory needs precision they did not supply. Whence, therefore, do explorers think the wild and ancient borders of mankind really began?

In passage down the hall of our forefathers every plinth is shrouded by a dense and theory-driven mist of guesses. In and out of this opacity drift phantoms; ghostly shapes arise in bloodless lines connected only by our ignorance. You might drill down using sharp technology and thence evolve full-colour illustrations of these busts like classic deities - for what are they but population of imagination?[80] And, as we recede, before what Royal Source, what Principal upon a throne, do we prostrate? We stumble on which Father of Mankind?

No-one knows the fractured Father Bone so take your pick of simian scraps. Or even, by Jove, an entire lemur. Was 'Ida', for example, the progenitor of lemurs, apes and men? In 2009 the well-preserved bones of a lemur-like primate, found in Germany in 1983, classified *Darwinius masillae* and dated 47 myr, were exposed to lucrative hype and, in the typical manner claims for missing links continually come and go, gross media manipulation. Overblown claims ('missing link found', 'eighth wonder of the world') and, launched with great fanfare, a publicity blitz ephemerally propelled the sexless fossil christened with a girl's name to an eager world. Even the normally circumspect Sir David Attenborough was sufficiently excited to declare prosimian Ida showed us who we were and where we came from; there was not, he claimed, a shred of doubt -

---

[80] *SAS* demonstrates the insubstantiality of materialism.

although both philosophically and factually there exists considerable bone of contention. Millions of specific, genetic slips would have had to occur between Ida (who lacks any anthropoid feature) and yourself and so the galvanic term 'missing link' here fizzles into something as blurred and scientifically meaningless as pointing to a lemur in a zoo. But that theoretical imperative drives heavy speculation. Are you, without a shred of doubt, overwhelmed by proof of such an 'obvious' piece of evidence? Experts sharply disagree. Ida's now 'down-graded from *the* common ancestor to some simian branch-line which perhaps just withered fruitlessly. The moral of this overheated tale must be, "Do not lean on lemurs over-heavily"!

Where, therefore, to start? The fossil record of extant great apes is poor. Enthusiasm might identify the modern as an ancient kind or, worse, construe what's ape as missing link. Of over 5000 species of ape about 4900 are deemed ancient and only about 120 are in circulation today. Whence, far out, did they emerge?

Some have speculated that a 40 cm stereoscopical 'gibbon' swung along the monkey-primate border long ago. Bjorn Kurten believed that the 'common ancestor' of both apes and men was a creature like **Propliopithecus** (~25 mya) of the Fayum in Egypt. Its molars match an australopithecine's and it is very similar to a few other ancient Egyptian bones dubbed **Aegyptopithecus.** *But* parallels have been drawn between both types and an extant primate, *Alouatta,* the howler monkey.[81] So perhaps, despite human molecular data relating closer to chimps and gorillas, gibbonish, European **Pliopithecus** (~12 myr), usurps the ancestral crown.

It has been noted that no fossils have been found unequivocally ancestral to chimp and gorilla but not man. Perhaps, therefore, in contrast to Kurten's fossil-based speculation, certain human genetic data more closely resembling gorilla than chimp sequences might be construed to suggest a common ancestor for the gorilla-chimp-man cluster (*Homininae*) at ~5 mya. A few have even speculated that *Australopithecus,* extinct for around a million years, is now represented by chimp and gorilla for which there are virtually no fossil ancestors. Brain sizes overlap; Zinjanthropus and Lucy's small, composite skull are morphologically similar. Do modern chimps therefore represent a *reversion* from a more man-like ancestor?[82] That such ideas can even be mooted[83] indicates the degree of flux in 1980's palaeoanthropology.

---

[81]   Fleagle J.G. 'Ape Limb Bone from the Oligocene of Egypt', *Science,* vol.189, no. 4197, 11-7-1975, p.136.

[82]   ~400 kya fossil chimp found in 2005.

[83]   Cherfas J. and Gribbin J. 'Descent of Man – or Ascent of Ape?': New Scientist vol. 91, 1269, 3-9-1981, ps. 592-5.

Kurten also argued that Punjabi ***Ramapithecus*** (~16.5 to 5.3 mya) was the late Miocene ancestor of *Homo sapiens*. This creature, a forest-dweller, was only four feet high and known, until 1975, by about forty teeth and jaw fragments. Was it really an early hominid? More complete specimens were found in 1975 and 1976, which showed that *Ramapithecus* was less human-like than had been thought. It began to look more like ***Sivapithecus*** (~12-8 mya). Could it be that *Ramapithecus* was just the female form of *Sivapithecus*? Recent expeditions to the sub-Himalayan Siwalik hills in Pakistan have uncovered more 'pre-human' fossils of the look-alike ramapithecine and sivapithecine varieties. A new skull, by far the most complete *Sivapithecus* yet found, reinforced the once-minority-but-now-majority view that these creatures were more like orang-utans than anything else.[84] They could be the orang-utan's granddad except that, in 2002, a sandpit worker in Thailand came across one's mandible (~8 myr). On the other hand, a baboon, *Theropithecus gelada,* which inhabits the mountains of Ethiopia, has been compared with *Ramapithecus.* It has incisors and canines which are small relative to those of extant African apes; also close-packed heavily worn cheek teeth, powerful masticatory muscles, a short deep face and other 'manlike' features possessed by *Ramapithecus.* But no one doubts that it is a baboon. Nor opines that Rama-Siva puzzle-pieces were our ancestors.

So did some other fragments called ***Proconsul*** (5 myr at latest) father chimps and us? That would leave a couple of million evolutionary years or so to randomly triple brain size and 'invade' human, *H. erectus* and Neanderthal zones. No doubt, there are many bones to find; one lives in hope but will they staunch confusion or cool hot contention down?

20.3   The Toumai Skull

Staunch confusion they will not. The Toumai Skull (***Sahelanthropus tchadensis***) was discovered by Ahounta Djimdoumalbaye in 2001 in Chad,

---

[84]   Andrews P. 'Humanoid Evolution', *Nature,* vol. 295, 21 January 1982, p. 185.

southern Sahara desert. Estimated age (difficult to be sure since the bone was found on a sandy surface) is between 6 and 7 mya. This is a distorted but mostly complete cranium with a chimp-sized brain (between 320 and 380 cc). It has many apelike features, such as the small brain-size, brow ridges and small canine teeth which are characteristic of 'later' hominids. Because anything as old as this is (inevitably in evolutionary terms) in line for ancestry, the chimp-like fragments have been associated with an LCA (last common ancestor) at or before the man-chimp divergence. Is Toumai an Australopithecine or from an extinct population of modern chimps? Some experts have even speculated the owners of such bones were interbreeding for another 2.5 myr (which brings them to the ~5 myr Sarich-estimated chimp-human split). Yet, in 2016, a hominid lower jawbone (Greece) and premolar tooth (Bulgaria) called ***Graecopithecus freybergi*** were dated at 7.2 mya; that is, out of Europe came a man predating African *Sahelanthropus*!

Forms come and go within an archetype's expressed plasticity.[85] The first chimp-like ***Dryopithecus*** (9-12 myr) was found at the site of Saint-Gaudens, Haute-Garonne, France, in 1856. Other *dryopithecids* have been found in Spain, Hungary and China. They were about 4 feet long and more closely resembled a monkey than a modern ape. The structure of mid-Miocene limbs and wrists show that it walked as does a modern chimpanzee. *The first actual chimpanzee fossil,*[86] consisting of a molar and two incisors, was found in the Eastern Rift Valley, Kenya, Africa in 2005; this example of Pan was chipped from the same layers as *H. rhodesiensis* and *H. erectus* and dated ~400 kyr.

20.4   *Oreopithecus bones and diorama*

Was Italian swinger, ***Oreopithecus bambolii*** (1872 and 1958; ~8 myr), our antecessor or just a three-foot tall Dryopithecine ape? An almost intact skeleton of the 'Abominable Coalman' was prised from blocks of coal mined from a depth of 200 metres.[87] The initial reconstruction of its short-faced skull considerably resembled Broken

---

[85] *SAS* Chapters 16 and 22.
[86] Nature 437; 1-9-2005 ps. 105-108.
[87] New Scientist: 228: 30-3-61 ps. 816-818.

Hill Man (and thus 'archaic' *sapiens* or *H. erectus*). Its teeth are small, hands ape-like and locomotion interpreted as able to swing through trees combined with an unusual, simian sort of 'facultative bipedalism'. Most experts agree that 'Hill Ape' is an extinct but enigmatic hominoid or perhaps even (according to analysis by Johannes Hürzeler) hominid. However, fresh analysis in the 1990's identified an ape-like Dryopithecine. There is, of course, still contention as regards the taxonomic status of such an interesting, 'mosaic' primate. Some have suggested it is a unique type of ape and many that it cannot, due to anomalous age, be an ancestral hominid.

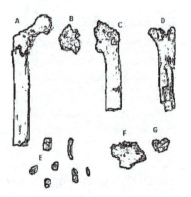 

20.5   *Orrorin tugenensis*

The search goes on. **Orrorin tugenensis** (~6 myr) was discovered in 2000. To some its handful of Rift Valley shards (including the posterior part of a *mandible* in two pieces, a few unattached teeth, human-like femoral fragments and some finger-bones) seemed to indicate it was bipedal and thence a step ahead of *Australopithecines*. However, the notion that it might therefore sideline the younger chaps from our ancestry was unpopular and itself soon sidelined.

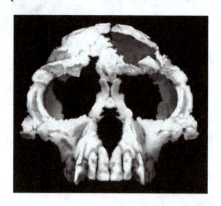

20.6   Reconstructed *Ardi*.

Not much younger! In 1992 Tim White discovered *Ardipithecus* in in the 'hominid-rich' land of Awash, Afar, Ethiopia. Two species have been delineated, *A. kadabba* (~5.6 myr) and *A. ramidus* (~4.4 myr). In the regular pattern of evolutionary pot-boilers 'Ardi' or *Australopithecus ramidus* was trumpeted as 'hardy-man', a missing link nearest chimp-man divergence that we shall ever find, a 'Rosetta stone for understanding bipedalism'[88] and so on. In fact, the skeleton was found in crumbling smithereens scattered in 17 locations over a distance of 2 kilometres and in need of extensive micro-CT scan reconstruction. A toe-bone critical to the hypothesis that Ardi 'almost certainly walked upright' (or perhaps, with 'facultative bipedality', could have walked upright as monkeys sometimes do) was found 15 kms away. Actually, the big-toe was graspingly suitable for swinging through trees and crushed pelvis shape unclear. Criticisms were made, Ardi reclassified to a new genus (*Ardipithecus*) and claims for direct-line ancestry to man quietly dropped.

Pin-head 'Ardi' (crushed skull reconstructed to ~350 cm³, that of a female chimp), was only briefly fêted as one of *our* grandmothers but was she progenitor of those extinct forms of chimp, *Australopithecines*? Or do the couple run too close in time? *Australopithecus anamensis* (or *Praeanthropus anamensis*) from the same area is dated ~4 myr. Almost a hundred *anamensis* fragments (e.g. KNM-KP 29281, jaw and chimp-like ear canal) representing over 20 individuals have been uplifted from the earths of Kenya and Ethiopia and raised to standing in the Human Hall of Ancestors.

Due to insufficient material researchers are not able to make enough observations to differentiate between many of the early hominids but do you remember that pesky femur (KNM-KP 271) from Kanapoi? It looked entirely human but, due to age (over 4 myr based on faunal correlation) and evolutionary theory couldn't be! So it was reassigned to *Australopithecus africanus* - until in 1994 Maeve Leakey's expedition found a fragment of human tibia (KNM-KP 29285) and quite promptly subsumed both into her newly-formed *anamensis* collection. She had uncovered several additional fragments of this hominid, including a *lower jaw* bone closely resembling that of a **common chimpanzee** *(Pan troglodytes)* but with smaller teeth more human in size. However, in 2006 more chimp-like specimens, this time with large instead of small teeth, were found a few miles from where Ardi died in Ethiopia. *Anamensis* upper post-cranial arms were simian but, due to the Kanapoi pair, it was judged 'habitually bipedal'. Would not human suggest obviously some element of humanity? If you classify what is human simian then, logically, you create an ape-man. You're free, unless it represents a dead-end branch, to tick it off the missing-link list.

---

[88] Tim White quoted in Science 295 15-2-2002 ps. 1214-19.

Of course, no evolutionist sees reclassifications and date-swapping as anything but trying to better fit his preconceptions. His intellectual frame is presumed completely true but therefore doesn't scientific objectivity become infused with subjective, evolutionary interpretations? Thus, due to proliferation of fossils and experts splitting these into larger and smaller 'ranks' of species, teasing out the single truth about an assumptive tree of life becomes complicated.

Or does it? Here comes **Lucy**.

In 1972 Frenchman Maurice Taieb took Johanson to the region which Teilhard had visited in 1928 when returning to China from France. Expeditions since 1972 have unearthed various fragments of 'a family of

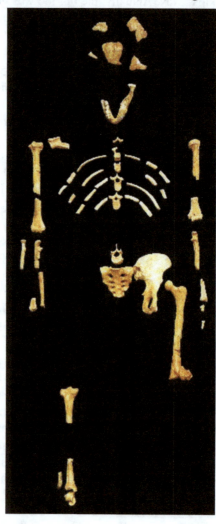

20.7 Lucy Neatly

hominids' said to have died in a flash-flood (but found in thirty-three feet of mudstone!). In 1974 Don Johanson's team brushed metre-high AL 288-1 (~3.2 myr old) from the dust of Afar, Ethiopia and classified her as a hominin, *Australopithecus afarensis*. This, though incomplete, is a skeleton to conjure with. Arrayed in portrait as if upright, she amounts to a third or less of a full chimpanzee-like skeleton. The arms are long and fingers curved as an orang-utan's. And the pelvis - which would be broad in a human - is said to have been distorted and to have undergone laboratory 'correction' to reduce this distortion. How can anyone know that the bones of a unique specimen are distorted? And by how much? How, moreover, will they recognize the 'right' position and will this become further evidence for Lucy's bipedalism? In fact, evidence from arm-leg ratios is inconclusive, not least due to the condition of these limbs. In 1973 two bones of a knee joint had been found in a stratum eighty metres lower than Lucy's. Although the distance between them and also their distance from the 'lady' is unclear it was assumed they were from the same individual and she is not actually a mixed-species composite! One (a tibia) shows, like Lucy, no human characteristics. The upper bone (femur), the very part which ought to show bipedalism if it existed, is badly crushed at the knee-joint. It was, nevertheless, due to interpretation of the way it may have attached to the pelvis, claimed to exhibit unquantified evidence of upright posture so that Lucy had 'spring in her walk'.

The girl's bipedalic spring was, however, later buffered. Her foot had been reconstructed from a collection of bones from several different species and its navicular was, as in our own case, arched. Where had this curvaceous navicular, vital for full bipedal locomotion, come from? William Harcourt-Smith and Charles Hilton challenged[89] that, although some of the bones were Lucy's (3.2 myr) others, including the navicular, were from individuals dated ~1.8 myr. In other words, the reconstruction was actually created from a mixture of bones from *A. afarensis* (3.2 myr) and *H. habilis* (1.8 myr). The crucial navicular, from which the arch of a bipedal foot is shaped, was of the later age. Was this mix-up careless or deliberate? Was Lucy really flat-footed with a simian gait? Or should *afarensis* be reassigned to earlier *anamensis* (for which no foot bones have been found)? This would make the very human footprints from Laetoli (now also *anamensis*) even earlier - worse still for evolution if they *are H. sapiens*.

Thus, from the doubtful witness of one small bone found eighty metres deeper and some distance away and a second from a different species, a star was conjured up. Lucy, an ape in skeletal respect, was

---

[89] Scientific American: August 2005 ps. 9-10.

welcomed to the club that leads to humankind.[90] How much imperative and skew were there in play?

Of course, contemporary chimps and other apes are happy in trees, running short distances on land or fording streams but are *not* fully bipedal. For example, rain-forest chimpanzees, *Pan paniscus*, spend a good deal of time walking upright and some apes hunt in part on foot but nobody calls them 'obligate bipedals', like humans. Is *afarensis*, like the Leakey team's distorted skull called *Kenyanthropus* (KNM-WT 40000, 3.3 myr, found in two places and sporting features of both *A. anamensis* and *H. habilis*), really just a kind of extinct chimp?

Johanson had at first classed Lucy as human in the genus *Homo* but by 1979 he had concluded that she was at the very base of human stock, pushing at the limits of hominid divergence from the apes. By everyone's admission the creature was 'very primitive' and its closest living representative would probably be the chimpanzee. The unbeliever, in spite of Johanson's conviction that this fragmentary, ape-like skull was perched on a bipedal body, would note the missing hands and feet and crushed knee-joint. He would suggest it was not human at all but a species representing a form much closer to chimpanzee or gorilla. A complete skull, made up from fragments of different individuals, looked like a small female gorilla. And the palate is considerably more like that of a chimpanzee than a human. Still, could it have been an ancestor? Why don't *Australopithecines* (or *erecti* and the rest) run free today?

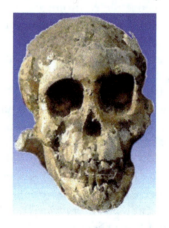

20.8   Lucy and mate   'Lucy's Baby' 3-year 'Selam'

These and other similar but incomplete discoveries have been eclipsed by another *Australopithecine* star, the recently raised '**Dikika girl**'. The incomplete fossil of this toddler, also named 'Selam', was exhumed from

---

[90]   New Scientist 97: 20-1-83 ps. 172-77.

near Lucy but, possibly being 100,000 years older (3.3 myr), is not really what her nickname, 'Lucy's baby', implies. Is she 'mankind's mother' either? She could not have told you so because the hyoid bone beneath her skull (comparable to an infant chimp's) barred speech. Her teeth and upper body shape including shoulder blades are simian; arms and hands are fine for knuckle-walking and for swinging through the trees, as is the ape-like inner ear for ape-like sense of balance. No pelvis is available but a partial foot and the angles of bones in Lucy's reconstructed and Dakika's unreconstructed hip suggest that *Australopithecine* 'chimps' might, like other chimps, have been able to walk upright if occasion required. In 1970 anatomist Sir Solly Zuckerman refuted Dart's claims for such a stance; and few anatomists support Johanson's claim that Lucy mostly roamed this way. In fact, detailed analysis has shown that *Australopithecines* are more different from humans and apes than the latter couple are from each other. Now, older than Lucy, over a hundred more fragile, crushed and widely dispersed bones have been reconvened into a composite creature from the same 'hominid-rich' region of Middle Awash, Afar. This is the way of simian reconstruction - unreconstructed aping of an academic line.

Back now but forward to Dart's baby (***Australopithecus africanus***)! We have begun to see that *Australopithecines* are not so much a grey area as one hidden from objective view by the blinding light of revelation after revelation. They are regarded variously as apes, ape-men or members of an extinct sideline which, although emergent from a common ancestor, were thereafter never a human ancestor. Dart, with Broom, thought his baby was our Great Grandparent; but Grafton Elliot Smith and most contemporary workers disagreed, concluding that the Taung fossil was essentially identical to the skull of an infant gorilla or chimpanzee. Other bones were collected from Kromdraai, Swartkrans and Makapansgat (also in South Africa). Professor Solly Zuckerman and experts on his staff investigated these bones for some years and came to the conclusion they were indeed apes.

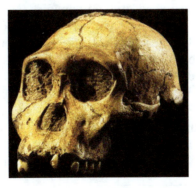

20.9 *A. Sediba.*

If you have to leap the *Australopithecine* to *Homo* gap you're aiming, hypothetically, to start a 'muddle in the middle'. This 'middle' is assigned, time-wise, a period between about 2.3 to 1.8 mya. Lee Burger, today's transition man, has enrolled at least one new member to his class. In 2008 a couple of partial skeletons called ***Australopithecus sediba*** were found at Malapa, also a 'Cradle of Mankind' site but this time near Sterkfontein in South Africa.[91] Has *sediba* knocked Lucy off her pedestal? The rounded skull (~420 cm³) of an adolescent is dated at ~1.95 myr. Long arms, teeth of disputed provenance, jaws, spine and knees possibly of *Homo* type and a simian heel have rendered this discovery 'mosaic'. Did s(he) leap *and* run? The heel-bone would have rendered bipedalism most uncomfortably twisted - unless the sample was a man's type injured, malformed or broken during its interment. We return next chapter (Muddle in the Middle) to mosaic *sediba*, this time in the company of oriental Flo.

20.10   Zinj-like KNM-ER 406 (gorilla?)

Could yet another youngster, at 1.75 mya, be Our Father? Olduvai Gorge in Tanzania had yielded up no hominids but Louis Leakey felt it would. In 1959 Mary, his wife, came across teeth and a shattered skull in association with both stone tools and the fossilized bones of many animals which had been broken open for their marrow. This led to a precipitate belief that the bones had once inhabited a tool-making hunter. Leakey honoured his 'star' with the first of its multiple stage-names - not *Homo* but ***Zinjanthropus***, 'Zinj' or 'East Africa Man'. Such 'paydirt' was treated to excited publicity by a sponsor, the National Geographic Society. Artists commissioned in the early sixties reconstructed 'identikit' portraits (see *Cartoons*: *fig.* 5.2) from the single skull - itself a reconstruction.

---

[91]   Science 9-4-10 ps. 195 *ff.*

Different hair styles were employed so that the 'man' ranged from apish to low-brow, even high-brow, human. Due to huge teeth 'he' was also dubbed 'Nutcracker Man' (although apes including gorillas use stones and not their teeth to crack nuts open). In the NG magazine Louis exclaimed that the find was 'quite obviously human'.[92] But given lack of post-cranial material, the size of its teeth, cranial morphology and capacity ($\sim$500 cm^3) and the large sagittal crest, a simian feature, he recanted. Nor was there proof this creature had ever manufactured or used the nearby stone tools; indeed, the eviscerated bones of prey suggests that 'Zinj' might have served as supper for a more advanced type. So the creature, catalogued OH 5, was classified as robust *Australopithecus boisei.*

A. boisei could not have been a missing link. Its time-line, from $\sim$2 to 1.5 mya, runs alongside *H. erecti* (such as Turkana Boy, Dmanisi skulls, Modjokerto infant, Java Man, KNM-ER 3228, Laetoli footprints, KNM-ER 1470 and so on). Therefore, how could it have given rise to these and thence *H. sapiens*? And it was found slightly *above* several more advanced and so-called habiline fragments (see next Chapter). So 'Zinj' was soon shunted off the human line into a siding. The down-graded genus was reclassified *Paranthropus boisei.* Similar species, *P. robustus, crassidens* and KNM-WT 17000 ($\sim$2.5 myr, also known as the Black Skull and classified *aethiopicus*) may also inhabit this cul-de-sac.

Anyway, that bony crest of calvarium undoubtedly bespeaks full-blown ape-hood so that by 1965 even Dr. Leakey had conceded that 'Zinj' was most like a gorilla (say, *Gorilla boisei*). But a non-retractive National Geographic did not fanfare this announcement!

Confusion over the evolutionary status of *Australopithecines* is illustrated in the case of Professor Ashley Montagu, a leading American anthropologist. In 1957 he wrote that such an extremely ape-like creature as Dart's baby could have nothing to do with man. In 1964, with Dr. G. Loring Brace, he changed his mind and placed *Australopithecus* in the genus *Homo.* But by 1977 he had reclassified it again as the genus *Australopithecus.* Such vacillation within an evolutionary mandate casts no doubt on the professional integrity of Dr Montagu; it simply highlights the insubstantial nature of evidence, imperative and skew.

Australopithecus, at four feet, never grew much larger than large chimpanzees. Dental analysis has shown that they were, like *Ramapithecus* from whom Kurten imagined they were descended,

---

[92] National Geographic Magazine: Sept. 1960 p. 421.

vegetarian. But if a group of them decided to swing down from the trees and become meat-eating *Homo erectus* on the plain, upright gait would be the last thing they would want. Their first efforts would give them an uncomfortable short-stretch roll, and a slow one at that. Man walks about as fast as a chicken; he runs upright at 12 m.p.h. while the paras monkey can run two-and-a-half times as fast. Indeed, the new man would have been about the slowest mammal on the savannah; rolling like a boat in high seas, and still wearing that tiny chimpish head, he'd have had little chance in the survival stakes. *Australopithecus,* if he wasn't some kind of chimp, didn't make it. After living side by side with modern man it is presumed to have become extinct about a million years ago.

Why don't we understand the relationship between these many species or find these any of them now? There are several answers, dependent on your perspective, to this question.

The standard *bottom-up* reply is that each one was, according to 'survival of the fittest', eliminated by a later, fitter model. In the case of 'phyletic gradualism' a species everywhere suffers replacement by a later 'conqueror'. This may also happen in the theory of 'punctuated equilibrium' wherein nothing changes except when some extraordinary stimulus drives rapidly; then the result may usurp its predecessor.

A more likely answer, a genetic one, involves alleles and adaptive potential (for both see Glossary). An *allele* is gene variety (say, for blue or brown eyes) and you partake of alleles from the whole human gene pool. From what genetic studies have been made it appears small populations were the order of the early primate day and, in this respect, genetic drift means that (especially in such populations) certain alleles may, by chance and over time, be bred out. In no way does this explain the informative origin of any allele (and its protein) but it explains its loss and thereby a changed population.

Adaptive potential involves changes to super-coded switches and recombinant refinements intrinsic in the genomic program of a particular type. In practice, the combination of epigenetic interaction with environment and genetic recombination may produce an intensive 'founder effect' when a small group is reproductively split (by whatever circumstance) from a larger main group. Obviously, its members take with them a reduced gene pool and we know (from the cases of Pitcairn and Faroe Islanders, Eskimos, Ashkenazi Jews etc.) that distinctive differences from a main group may accrue. Such a process of adaptive radiation may produce what was referred to in Chapter 9 as 'allopatric speciation'; extending to the minor level of what is termed 'micro-evolution' it occurs quite frequently and with all types of organism. It is also acceptable, in a secondary and random mode, to a *top-down* explanation of how extant and extinct chimp and human forms can morph.

Beak, fur colour and other genetic and epigenetic changes can occur rapidly and as a result of environmental pressures. Why should teeth size, cranial morphology, body-build and so on not (within the same strict typical parameters) do so as well? *Called 'adaptive potential'[93] the informative software for such flexible response is inlaid in the original genome and tapped using (as do computer programs) a multitude of switches and operating mechanisms (such as splicing). Such adaptation is not a matter of chance and the Darwinian explanation (mutation) is, except when describing damage, rendered redundant.* **From a *top-down* perspective this is the primary method by which chimpanzee, human and any other type of organism varies over place and time. You would expect the fossils to record it.**

They do. Before they died out, the *Australopithecines* never really changed in three million years. Their brains never evolved in size. Johanson writes that, by Richard Leakey's definition, modern chimpanzees would be classified as *A. africanus*.[94] A variety, *A. robustus,* has more massive teeth and jaws and possesses bony ridges (sagittal crests) found in gorillas and orang-utans. Was its gracile form the progenitor of *Habilines* (next chapter) or simply the sexually dimorphic, female form of *robustus*? Were both, as has been suggested, mates of *A. africanus*? Endless speculation engulfs old primate bones. **For the teleologist, australopithecines are the stuff that dreams, fame and fortunes are made on. In fact, they compose different extinct varieties, species and maybe even genera of chimps and possibly other apes.**

Are any ape-forms showing signs of incidental, nascent evolution in our day? Why not? Why don't we somewhere spot another human evolution, a second natural transformation now ascending? Because, it's argued, accidents don't happen repetitiously? Who, though, can logically counter a species of argument wholly reliant on the notion 'anything may or may not happen', that is, where every move is based on chance? *And so, because of this, can evolution ever logically, scientifically be proved or disproved?* **If not, it's not a scienfific but a pseudoscientific theory.**

---

[93] *SAS* Chapter 23.

[94] Johanson, D., *Science,* vol. 207, 7 March 1980, p. 1105.

21. *Muddle in the Middle*

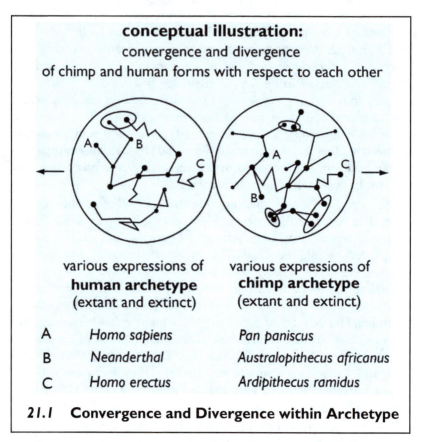

conceptual illustration:
convergence and divergence
of chimp and human forms with respect to each other

various expressions of **human archetype** (extant and extinct)	various expressions of **chimp archetype** (extant and extinct)
A *Homo sapiens*	*Pan paniscus*
B *Neanderthal*	*Australopithecus africanus*
C *Homo erectus*	*Ardipithecus ramidus*

21.1   Convergence and Divergence within Archetype

If it is ape | man there'll be no muddle in the middle, only (as the *top-down*[95] illustration shows), variation within the macro-boundary of an archetype. These variations may converge or diverge with respect to a neighbouring archetype (e.g. *H. erectus* may be considered of more simian feature than *H. sapiens,* that is, more convergent upon the chimpanzee archetype). **As opposed to ape > man macro-evolution a strong case can be made, as preceding chapters have shown, for variation within simian and human archetypes.**

If, however, it is ape > man there will be muddle in the middle. This is because there needs, theoretically and therefore actually, be a line of intermediate forms that bridge the ape > man gap.

---

[95]   for anti-parallel *top-down/ bottom-up* perspectives see 'For Starters'; also *SAS* Chapter 0: Two Pillars - a Dialogue of Faith.

The theory of evolution involves a strong urge, a primal instinct to arrange in series. Howell's parade (see *Cartoons*) is a typical expression of such overwhelming drive, such necessity that alone selects and thus ensures survival of its materialistic paradigm. *A corollary of this urge involves the classification of as many as reasonably possible so-called intermediate genera or species.* Thus, as regards the materialistic *aka* scientific study of human origins, a profusion of classifications is, we've seen, devised to represent evolutionary steps. Practically every new-found fossil fragment, often 'the most important solution to problems of primate phylogeny', is assigned an authoritative name to add to an impressive list. Such authority is then employed to promote the theory as a whole.

You might want to minimise the ape > man gap but here exist **two problems**. First let reptiles, for example, speak. Two practically identical alligator skulls, one fossilised (~75 myr) and the other contemporary, have, according to palaeontological tendency, been assigned a different genus and thus different names (*Albertochampsa langstoni* and *Alligator mississippiensis*). This, subtly but perhaps deceptively, lends variation evolutionary credibility.

The second self-deceit derives from exclusivity. Since archetypal separation is forbidden as a heresy, evolutionary imperative luxuriates in speculative family trees. These trees (or their bushy branches) sprout an ever-changing foliage. Bewildering bushes of hypothesis, dotted line phylogenies and, in a far-from-clear terrain, a lot of questions are marked up. It keeps the business busy but still, at this point, you might well interpret all the range of bones as ape <u>or</u> man without much substance only yawning gap between. In other words, there is, for a humanistic scientist, necessary consensus over evolution but seemingly endless, molecular and fossil-fuelled debate concerning how its macro-evolution *must* have somehow worked.

What, however, if the whole exercise is philosophically flawed from its first non-holistic, *bottom-up* step?

To facilitate the serial transformation, ape > man, several features are identified. These, which conversely might also obstruct, include cranial, ribcage, pelvic, femoral, hand and feet morphologies; also an increase in brain size and leg length, a decrease in teeth size, chin size and arm-length, and a smaller, less prognathic (pushed out) jaw. Of diagnostics, though, the key is bipedalism. **Bipedalism is considered the most important single event in human evolution.** In some scenarios it is even thought to have promoted brain enlargement (greater intelligence) and complex tool-making. *The second corollary of primal evolutionary urge is, therefore, one we have already met - the identification of possible bipedalism.* In the cases of, for example,

139

the Laetoli and Koobi Fora footprints, Olduvai hominids (e.g. OH 62 found a few yards from the Laetoli impressions) and 'Lucy' (AL-288-1) the issue of bipedalism's first moment has been contentious to the point of straining truth in favour of an imperative, that is, an interpretation that conforms with evolutionary expectations.

In his book, *Origins,* Richard Leakey refers to "...the Leakey tradition, that you look and look again until you find what you know must be there."[96] It is difficult to imagine a more dubious philosophy for an impartial scientific worker to adopt. Some evolutionists may share it, though perhaps they are not all clear enough in their minds to say so. *However, if there is just one thing that every evolutionist agrees on, it is that evolution happened: it's just a matter of looking until you find the evidence.*

"Perhaps more than any ether science human prehistory is a highly personalized pursuit, the whole atmosphere reverberating with the repeated collisions of oversized egos. The reasons are not difficult to discover. For a start the topic under scrutiny - human origins - is highly emotional; and there are reputations to be made and public acclaim to be savoured for people who unearth ever older putative human ancestors. But the major problem has been the pitifully small number of hominid fossils on which prehistorians exercise their imaginative talents."[97]

So wrote Roger Lewin in an article about the opening of The International Louis Leakey Memorial Institute for African Prehistory (TILLMIAP) in Nairobi. The trouble is simple enough: there is not a great deal to work on. As we have seen, the remains consist mostly of fragments of jaws, broken skull pieces or feet. Not even half an ape-man skeleton exists to prove that man was beast. Yet the more fragmentary the remains, the more sweeping the claim, the more presumptuous the presentation of the evidence and the brighter the blaze of publicity.

There's no clear middleman. For evolution's ape > man it's all a moving show. Yet in the narrative still apes and humans stand apart. Milford Wolpoff[98] summed the situation up. *'The phylogenetic outlook suggests that if there weren't a Homo habilis we would have to invent one.'* The transformist trick is taking two like forms and then turning (or evolving) one into the other. Which, therefore, are the chosen stepping-stones, the serial set of forms that have to pass us to our human side?

---

[96]   Leakey R. and Lewin R. *Origins,* MacDonald and Jane's, 1977, p. 106.

[97]   Lewin R. *A New Focus for African Prehistory, New Scientist,* vol. 75, no. 107, 29 September 1977, p. 793.

[98]   American Journal of Physical Anthropology; Nov. 1991 p. 402).

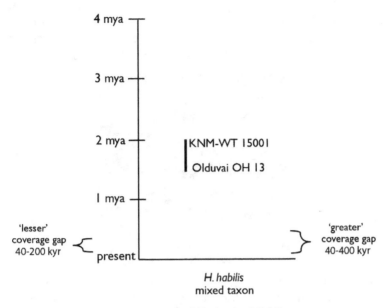

4 mya

3 mya

2 mya — | KNM-WT 15001

Olduvai OH 13

1 mya

'lesser'
coverage gap
40-200 kyr

present

'greater'
coverage gap
40-400 kyr

H. habilis
mixed taxon

21.2   Muddle in the Middle

On cue, therefore, enter **H. habilis** to bridge a yawning ape > man gap. In the 1960's members of the Leakey family working at Olduvai Gorge in Tanzania unearthed cranium, mandible, foot-bones (the human-looking OH 8 from a couple of feet below 'Zinj') and hand-bones. The latter were interpreted to indicate manipulative ability and the foot-bones bipedalism. George (OH 16), Twiggy (OH 24), Cinderella (OH 13) and other habiline assignations enlivened with such catchy nicknames soon followed. Louis Leakey assumed that, as with 'tool-maker Zinj', the bones had once manufactured nearby tools and were not prey like other nearby animals. So he bundled the material into a collection he called 'Handy Man'. In 1964 *Homo habilis* was officially announced[99] to the world as the genetic and morphological middleman and thereby vital missing link that crossed to our own species. Since then these dry old bones have kicked up clouds of dust because 'Handy Man' is, while in theory a necessity, in practice subject of debate.

Who exactly was a *habiline*? Did s(he) derive from the *Australopithecines*? From a couple of cranial fragments the reconstructed capacity of OH 7 (the type specimen dubbed 'Johnny's Child') has been variously estimated between 580 and 710 cm³. Did *Homo erectus* and thence *sapiens* really ensue?

The Gorge at Olduvai is divided into seven main beds. The most recent (top) one, in which *Homo sapiens* remains have been found, and

---

[99]   Nature 202: 4-4-64 ps. 7-9.

the next two do not concern us. In Beds IV to II were *erectus* fossils and in II and I *Paranthropus*. Allegedly (because no-one has clearly defined their traits) *habilines* were only held in the lowest (I).

21.3   OH (Olduvai habiline hominid) 24 'Twiggy': skull crushed flat when found; fragments reconstructed using plaster and informed guesswork.

Several skull fragments (including OH 16 trampled into hundreds of pieces by cattle and crushed-flat OH 24), teeth and stone tools such as choppers used today in nearby Turkana were found in the lowest two beds at Olduvai gorge along with what was interpreted as a bone tool. At the bottom of Bed I (at ~1.9 myr) a jaw was found and, a mile away at the same level, a rough circle of loosely piled stones construed as a windbreak, a temporary shelter such as hunters make today. Who would think of a circular stone lodge as the work of apes? Rather, you might conclude, all this simply shows men lived with apes and ate their bush-meat just as now.

21.4   *Homo habilis*? KNM-ER 1470 and 1813 est. 1.8 myr (see also Chapter 'Broken Hill Man' *figs*. 19.8 and 9).

In 1986 Tim White and Don Johanson discovered a partial skeleton (OH 62, with similar cranium and teeth to Swartzkrans SK 847, Sterkfontein Stw 53, KNM-ER 1805 and 1813). These post-cranial remains yield a height just over an *Australopithecine* gauge of three feet. The organism was identified as *habiline* by its maxilla (upper mouth bone) but this and other diagnostics for both OH 62 and the species as a whole are, being unclear, problematic. In fact, as well as disagreeing on trait definitions, the experts argue which of the several dozen habiline candidates actually belong to the rustled-up, mish-mash species. The taxon appears to include both simian and human material. And no two researchers appear to attribute the same specimens to *habilis*.

Even Leakey did not believe that *Homo habilis* could have been a transitional form between *Australopithecus* and alleged *Homo erectus.* Various estimates place the earliest Australopithecines living over 4 mya and the latest (*Paranthropus robustus* and *boisei*) becoming extinct ~1.2 mya. *Homo erectus* is said to have inhabited an era ranging from ~1.95 mya (KNM-ER 3228 femur, KNM-ER 2598 cranial fragment, Swartkrans SK 27 and 847 and also specimens from Java, China and Africa) to 40 kya (Java Solo) and, arguably, as little as 6 kya (Australia). It has also been asserted that *sapiens* co-existed along the whole habiline era. This assessment means (where all dates are uncertain and to more or less extent arguable) that *Australopithecines* co-existed for over half a million years in the same area of Africa before competition for survival with *habilines* allegedly wiped them out. Was such protracted 'struggle' one of zero fierceness, that is, only theoretical? Even if *Australopithecines* had split into *Paranthropus* and *habiline* taxa about 2 mya it gave the latter precious little time to transform from three-foot apes into five or six-foot and essentially human *erecti*. Not only this, but *Homo heidelbergensis* is on the move by 800 kya (Gran Dolina, Atapuerca, Spain and Ceprano, Italy).[100]

Other critics query how (when thinner skull-bone is presumed evolved from thicker) thin-boned *habilines* could have reverse-evolved into thick-skulled *erecti* and *Neanderthals* before again thinning to human form. Variable *habiline* time-lines are proposed but all fall within ~2.1 myr (the cranial fragment KNM-WT 15001) to ~1.2 myr - about a million years. Consider now that a chimp is calculated (according to the mode of *DNA* analysis) to be between 92 and 98% genetically human; and that the LCA (last common ancestor) of chimp

---

[100] It is debated whether the fragments are from *H. erectus, heidelbergensis, neanderthalensis,* perhaps even *'archaic' sapiens* or two new species, *antecessor* (Atapuerca) and *cepranensis* (Ceprano).

and man is assumed to have lived ~6 mya. **The least genetic difference, 2% of 3 billion human base pairs, is 60 million. So we are expected to believe that over 6 million years a steady average of 10 beneficial but accidental mutations per annum have gradually, progressively transformed a certain line of chimps to you.** Of course, this involves purely standard evolutionary speculation, vast and unprovable but 'factual' because it's been decided that the theory's true. Thus, in the 4 myr before habilines (at 2myr) about 40 million beneficial changes will have occurred. Then, during the million years of *habiline* transition, a proportionate 10 million more 'beneficials' must have taken place in order to reach *erectus*; and next perhaps a further 5 million to reach Neanderthal beginning ~500 kyr. Now only another 5 million beneficial indels will reach *H. sapiens*! But if *erectus* was alive by ~2 mya there will have been no time for change from habilines! And if humans were alive that long ago the case becomes much worse! Should we not re-think? To make sense we might even need a shift from evolutionary paradigm.

21.5 The long and the short of it.

Such suggestion might provoke emotional outrage and scorn but now at least you see why there are almost as many ape > man genealogies as experts. Guess-work in order to fit facts to harder-than-fact theory is the order of palaeontology's day. *Today's research may be more honest, yet it is as strongly motivated as ever by a wish to validate theory, to render evolution proven in the fossil finds.* **In hardest fact, the whole *habiline* business is reminiscent of Piltdown not because of fake material but dubious assignment at a critical juncture (bridging the large gap between *australopithecine* and *erectus*).**

What, in the future, would palaeontology make of a fossil of the tallest adult male in the world (Sultan Kösen at 2.51 m) and, nearby, the world's smallest (Chandra Bahadur Dangi at 0.54 m)? What of their relative cranial capacities, teeth and bone size or any form of osteopathology? What of micro-encephalic (tiny-headed) and megaloencephalic (big-headed) individuals? Or disproportionate dwarfism that could, in a small and isolated group, be common? Would such immense but by-now prehistoric ranges be awarded different palaeo-species names? In what order did the long and short evolve?

21.6   Flo (chimp-like brain volume, ~ 370 cm³, and body length 1m.)

Let's for a moment slip across from West to East and inspect a case of 'midgets'. Said to resemble *Homo erectus*, the possibly female 'Flo' (or, scientifically, LB1) was discovered in 2003 at Liang Bua cave on Flores Island, Indonesia in the company of cooking kit, hunting gear, extinct dwarf elephant (*Stegodon*) bones and stone tools dated at 800 kyr. If not herself bush-meat she may have used utensils and fire. The chimp-sized skull was slightly thick-boned with trace brow ridges above eye sockets only, chin recessive and cranium low. Face and teeth were human-like but height and arm/ leg proportions (long arms and short, stocky legs) simian. The upper arm (humerus) was human-like but the legs may have been differently oriented to ours (from disease, dwarfism or normal anatomy?) so that any poor running would have had to be jogged along, as with modern apes, by knuckles.

She was predictably claimed as a new species (***Homo floresiensis***) by her discoverer and, equally predictably, 'aped' by National Geographic and other imaginative illustrators. But was 'Flo' (*aka* 'The Hobbit') of genus *Homo* and thus, as her given name dictates, human?

We know brain size is not, by itself, an indicator of intelligence. Why should allopatric speciation not have generated an isolated, small-bodied (and proportionally small-brained) race of mankind? Although even smaller than *Australopithecines*, this woman (or ape-woman) lived only 12-15 kya - young enough to be contemporary with modern man and, if human, to resemble modern wives, mothers and sisters. Is the pin-head 'hobbit' a missing link, a separate species of human or, as Australian palaeopathologist Maciej Henneberg claims from photographs (having been denied access to examine the actual skull), was one of her molar teeth filled by a dentist in the 1930's?! Seven more non-fossilised skeletons were found on site and assigned ages up to 95 kya.

An erstwhile student of von Koenigswald and doyen of local palaeoanthropology, Teuku Jacob, whisked away material, allegedly damaging some of it before return. He claimed Flo was 'a modern woman with a congenital disease that caused a shrunken brain-case' (a condition called microencephaly). In other words, she was a pathological specimen of *H. sapiens*. However, while hundreds of medical aberrations can cause microencephaly, dwarfism, cretinism and so on has even such non-evolutionary diagnosis proved necessary? Are not pygmies sprinkled worldwide? Teuku pointed out a community of over seventy families of 'little people' not much larger than Flo living at Rampasasa, an Indonesian village just down the road from her Liang Bua cave; and over twenty skulls of dwarfs or pygmies (with human jaws) were found on the Micronesian island of Palau about a thousand miles away to the north. Did her people therefore drift with ocean currents from here or Sulawesi? Or, confounding palaeoanthropological theory, did they invade from the west by boat?

H. luzonensis, consisting of 7 teeth, 2 finger bones 3 foot bones and a thigh fragment excavated over the last decade from Callao Cave on Luzon Island in the Philippines, underlines the phylogenetic problems. With an estimated height of four feet, no skull and a bone set not exactly like any other hominin, was this another oriental dwarf (assigned to genus *Homo*) or an ape of some variety? On the presumption that ape-human transformation is a reality, these bones do not support a smooth, unambiguous transformation. Nor has resolution of this twist in the monkey tale been helped by an absence of *DNA* sample.

The real question is whether Flo or Philippino fragments have dislodged *habilis* and Lucy off transformist perches. Adult humans vary in size but no-one's *DNA* survives long in the tropics so that perhaps, in this case, missionary Darwinism misconstrues the status of a modern human being. Since it's difficult to draw sharp, evolutionary branches, do the body in Liang Bua and bones in Callaoa represent muddle-in-the-middle *habiline, australopithecine* or poor and perhaps sick humans?

Africa is not an island archipelago but still palaeoanthropology craves to find, in a fossil record that is frustratingly ambiguous, unambiguous examples of an ape > man missing link. Might these render Howell's cartoon a clear, correctly time-lined illustration of transitional forms?

At another 'Cradle of Humankind', a World Heritage site of Maropeng near Sterkfontein in South Africa, Lee Berger of *sediba* fame, funded by the University of Witwatersrand and the National Geographic Magazine, fanfared another transitional species, ***Homo naledi***.

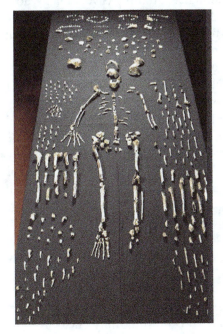

21.7 Homo naledi composite

Over 1500 fragments of at least 15 individuals were found on the floor of Rising Star Chamber and exhumed into daylight. The path to this 'crypt' involves a 100 metre exercise including Superman's Crawl through a slot only 18 cm wide followed by a sheer drop of 12 metres to the fossil site. It is claimed the pieces are 'homogeneous' i.e. from the same species - if not, how would you guess what fitted what? Reconstructive composition could produce mosaic combinations without check.

Skull reconstruction has indicated cranial capacity of between 450 (female?) and 550 cm³ (male?). Such capacity (less than half human) freed NG to produce, with all the sensationalism of the press, a fierce-frowning ape-man splashed above the banner 'Almost Human'! One partial skullcap is low and sloping, the other (comparable to *habilis* and *erectus* morphology) curved with more pronounced brow ridges. Teeth are small except an 'australopithecine' third molar. Upper body including long, curved fingers is 'primitive' (= simian) but a few traits mix things up. Wrist and palm bone composites are deemed more man-like; so is leg, foot and ankle reconstruction. The elements sum to a creature height of somewhat above four feet and include an interpretation of the potential for bipedalism.

The bones were strewn on the floor. They needed bagging not digging. There were no signs of butchery or predation; nor any animals, plant materials or stone tools commingled. And the cave geology does not *appear* to have changed over 3 myr (the date first assigned to the fragments). If it has not, how did the bones get slung collectedly down tortuous passages so far from its entrance? Was the orifice a dustbin or an unsophisticated graveyard? If either concept, hygiene or symbolic ritual, was involved then Berger claims *naledi* must have been almost as human as Neanderthal or *sapiens*.

More recently an almost complete skeleton (affectionately dubbed 'Neo') and more bones have been found. The arms of Neo's ape-like form (perhaps 40 kg and 1.4m high) are tipped with relatively humanoid hands. Dated between 230 and 330 ky, give or take a year or so, the age of the Star Chamber fossils has thus been severely reduced. In this case, since humans are supposed to have evolved ~2 mya, they may represent a relic, a line persistent long after branching to human stock; but there is no evidence, beyond Darwinian narrative, for any such putative ancestral relationship. As Chris Stringer once wrote; "The study of human origins seems to be a field in which each discovery raises the debate to a more sophisticated level of uncertainty". [101]

Is, therefore, Berger's judgment justified? Or is there more than this persuasion's tale has so far told? A campfire, flinty clicks as tools are chipped, the meal or, as they trek to hunt, a silhouette of ape-like pygmies in the human dawn. It is easy to clothe insubstantial ape-man in a fragile yet elastic web of story-telling. Socio- and psycho-biology of prehistoric family life and loves have, as a free imagination zone, evolved and still evolve in prolix profusion from a few stones and bones. In the *naledi* case of apparently deliberate deposition there is neither sign of ritual nor other entrance - even if long blocked.

---

[101] Scientific American May 1993 p. 138.

So has this creature joined the dotted lines that evolutionary theory demands? It is a problem that the finds have not been dated. Stone (hominin) tools from 3.3 mya have been found near Lake Turkana in East Africa. Did such pre-habilines lope south as far as Sterkfontein? Some experts suggest that *naledi* may have been an isolated, unique species, an extinct ape that's not a 'missing link'; and while others perceive Lucy-like features yet others reckon the cranial relicts may be more *erectoid* (for example, like the Dmanisi gang) than *habiline*. Maybe, therefore, over 3 mya, *naledi* lived in Lucy's time; or, at 2 mya, was an *australopithecine* preceding *habilines*; or, less than a million years ago, roamed *ergaster*-like. Perhaps these fragments even trespassed on the human era - if you're thinking in the only naturalistic way to read life's history?

They were not *habilines* or missing links but in 1929 Robert Yerkes[102] mentioned 'unclassifiable apes'. For example, 'Koolokamba' and, in Dresden zoo, 'Mafuka' seemed to show hybrid chimp-gorilla characters. Could transgenic mating ever happen in the wild? Some agree that 'pseudo-gorillas' might exist but others doubt it. Certainly and fortunately Stalin's 'scientific' attempts to force human-ape matings failed; and attempts to produce such hybrids by artificial insemination are outlawed. *Top-down*, it would seem more likely that, within a small and scattered population of ancient primates, isolation, reduced gene pool, allopatric speciation and adaptive potential could generate such variation as is shaved by palaeoanthropologists into ever more chips off old blocks, that is, new species.

Type, man or ape, is still debated. Should *naledi* not be *A. naledi* and *sediba H. sediba*? Has *naledi* usurped Lucy, *habilis* and Flo to take the top transitional place? And is *Homo naledi*, any more than *A. sediba* or *H. floresiensis*, really man? What a tangled tale that facts can weave! **And, for sure, if you believe in (human) evolution *H. naledi* might well validate your creed**. What, to reasonable men, is the alternative? Isn't it outrageously regressive, from a bottom-up perspective, to deny the mindless and unguided bio-universal process?

If, however, reason logic and mathematics aren't material; if there's an immaterial element called information; and if mind, symbolic thought and consciousness exemplify its natural, elemental power and potential then, top-down, the evolutionary paradigm (and its human fraction) might turn out the greatest of illusions.[103] **In this case, there is no middle to have muddle in. If man | ape then *naledi*, either way, cannot affect your creed.**

---

[102]   in 'A Study in Anthropoid Life'.

[103]   for systematic treatment of the issue see *SAS passim*.

22. Molecules (Not Fossils)

From bones to biochemistry. From anatomical to molecular taxonomy. This, as palaeoanthropology enters today's age of information, is the moment to rehearse some common sense. Life is built on information; it is developed from symbols that encode purpose in the form of programs. Such information is carried, for its ink and paper, on a complex molecule called *DNA*. *What is information? Where does it come from? Can unguided, natural forces generate the message that is needed to construct the simplest cell? In other words, in the beginning was there information or not?* **Materialism says no; holism, in a logically formulated way, includes its immaterial element.**[104] **It, therefore, says yes.**

Fortunately or unfortunately the materialistic community (including the physical and bio-sciences with attendant media) have made a philosophical decision 'no'. Information must, therefore, without reason and perchance evolve itself. *If this first step is wrong then a theory in no part more replete with speculation than in the field of 'human evolution' may also be, simply, 'barking up the wrong tree'.* **The *basis* of biology is information.** *N*evertheless, without full consideration of its possible origin, a naturalistic establishment and its various institutions will resolutely not consider the *top-down* philosophical alternative to its *bottom-up* choice. It is absolutely in denial. <u>**Such concrete closed-mindedness is ironic for a discipline that vaunts its open-mindedness!**</u>

This general closed-mindedness blocks despite the fact that materialistic science has no way solved the information enigma; despite the fact the random mutations over three and a half billion years could hardly come up with code for a single specific protein of 150 amino acids;[105] and despite its historical analysis having used the same not-strictly-scientific method of abductive reason (or forensic-type inference as to best explanation) as Sherlock Holmes. It blocks and thereby flippantly rejects an identification of a hidden factor, mind, as the present operation best generating information.

Such serious failure of common sense is not anyone's good news.

However, as regards its explanation of ape > man (or any other case of evolution) *bottom-up* Darwinian evolution is, past the fact of variation-within-type (*fig.* 9.2), shot through with speculation. *It assigns the reason*

---

[104]   *SAS passim*; especially Chapters 5-6 and, as regards life, 19-25.

[105]   see Glossary: combinatorial space; also *SAS* Chapter 23: Entropy of Information.

for remodelling of a body as genetic mutation, natural selection[106] *and local circumstance.* This process must be slow. Hypotheses are flexible as dates uncertain. Time is dispensed like money by a billionaire. Every fact is framed within this paradigm.

Top-down concentrates on information, program and their origin. **Its primary key is the pro-active flexibility of adaptive potential embedded into genome (program) by its switching sub-routines, splicing triggers, translation mechanisms, epigenetics and response to operators within (and perhaps circumstantially, from outside) the cell.** Of course, natural selection in local circumstance occurs but genetic mutation is rated a secondary post-active, random effect; it may hit a lucky chance survival but is largely neutral if not deleterious in its consequence. Who, after all, would improve a highly coordinated, regulated technical cynosure by scattergun exchanges born of chance? Thus, *top-down*, every fact is framed within an information-rich, holistic paradigm.

There are three main types of genetic feature we shall now consider - **autosomal *DNA*** (the nuclear genome less Y chromosome), **heterosomal Y chromosome *DNA*** (from which male lineage may be traced as if by patriarchal marriage surname) and **non-nuclear mitochondrial mt-*DNA*** (from which female lineage may be traced as if, were it the case, by matriarchal marriage surname).

Autosomal *DNA*

Autosomal *DNA* is the name for the nuclear genome less the heterosomal sex chromosomes, in primates X and Y. About 2% of its content is involved in coding for protein and the rest with regulatory, structural and as-yet unknown function; it is expected some small proportion will turn out, by entropy of information, to involve random mutation (or 'gibberish') of varying impact.

Genetic variation occurs, as far as evolution is concerned, by means of copying errors, damage or the uptake of foreign *DNA* (say, from viral infection). Such errors include insertions, deletions (called together indels), swaps of sequences from one place to another, duplication and simple substitution of one nucleotide of length of code for another. *Bottom-up*, such accidents drive variation and, for the theory of evolution, comprise the engine of unlimited innovation and adaptive plasticity of form.

It is generally agreed that the closest relative to man is chimpanzee. And by using comparative *DNA* and protein sequence data fed into the formulaic assumptions of a so-called 'molecular clock', the LCA (last common ancestor) of chimp and man is calculated to have lived ~6 mya. This date, whether right or wrong, therefore constitutes the point

---

[106] see *SAS* Chapter 22: The Editor.

against which all timing as regards 'human evolution' is calibrated. No paradigm other than evolutionary is tolerated and all quest for scientific understanding falls within such calibration's frame.

Humans are (including male/ female differences) 99.5% or more genetically identical; and Neanderthal *DNA* is upwards of 99% the same as ours. Meanwhile, all (phenotypic) variation in dogs is accounted for by 0.15% of their genome; and two species of mice (*Mus musculus* and *Mus spretus*) show the same genetic variation as between chimps and ourselves. Clearly, the genetic and structural components of an organism are not directly related. However, like a cat or mouse, chimps have a very great number of bodily similarities to humans and you would expect, whether by design or not, similar code for similar engineering.

Having said that, since 2005 much more chimp and human genome sequence data has become available.[107] Previously chimp/ human genetic difference had been estimated at between 2 and 8% depending on method of analysis e.g. size of *DNA* fraction, which parts of the genome are picked (sometimes 'cherry-picked) for comparison or the particular choices of *DNA* or amino acid sample sequences used to make comparison. Indeed, a mantric 98+% similarity used to be repeatedly touted. Now it appears likely that the mantra is widely inaccurate on several counts. Firstly, substitutions (point mutations) may account for over 1% of genome difference. Next add ~4% due to small-scale indels[108] and nearly 3% due to large-scale ones (including one chromosome fusion and major inversions in nine others).[109] Duplication of certain mobile elements (called *Alus*) and repetitive micro-satellite *DNA* accounts for a further ~6%.[110] By now similarity is reduced, as the journals indicate, to below 90%. Even more brazen, research by geneticist Jeffrey Tomkins perhaps drives it below 80%![111] Maximum identity is capped at 86-89%. **This sets an acute and inescapable trap for the evolutionist. Why?**

In sheer length a chimpanzee genome (just under 3 billion base pairs) differs in size from a human (~3.2 billion) by ~200 million base pairs. So either chimps lost or humans gained a lot of base pair 'letters' in their text. Let us allow a generously low 10% difference in make-up. This means ~320 million 'suitably-directional' changes towards the transformation of an ancestral ape (LCA) into a modern chimp and man - let alone the much greater quantity of 'negative' mutations or the genetic drift that would tend to lose either kind before they became

---

[107] Nature 437: 1-9-2005 ps. 69-87.
[108] PNAS vol. 99, 21: 15-10-2002 ps. 13633-35; for 'indel' see Glossary.
[109] Nature 437: 1-9-2005 ps. 88-93.
[110] Genome Research 14: 2004 ps. 1068-75.
[111] Answers Research Journal 4: 28-12-2011 ps. 233-241.

'fixed' in a genome. *320 million 'beneficial mutations' (called BMs) in 6 myr requires on average one 'positive' per couple of weeks fixed in the sex cells of each evolving germ line! Such speculative reliance on the unseen power of genetic randomness is statistically, let alone biologically, outrageously wishful 'science'.* **There is far too little time.** *At this rate man could not have happened.*

Let us, therefore, accord double generosity to the evolutionary camp. Take a mere 5% difference (95% similarity) amounting to some 160,000,000 changes. **This time simple calculation shows that, for 160,000,000 'correct' nucleotide changes to have occurred in 6,000,000 years on average at least one advantageous mutation must have accrued in gametes (as opposed to other body cells) every couple of weeks!** If this number represents the current chasm between chimp and man it may be allowed that half this number occurred in *each* species since LCA but, equally, re-emphasised that those mutations germane to the production of *Homo sapiens* would amount to a tiny fraction of those actually happening. *This is because the great but incalculable majority would be neutral, negative or 'reversing advantageous' in character.* By these facts and derived calculations ape > man evolution is like wishing on a star. It is effectively impossible. **Humans have not have happened in this fundamentally aimless way. Some other device would have to be ascribed.**

It is claimed that 30% of human protein is identical to the chimp homologue - unsurprising given that a much greater percentage of physiological activities are just the same. But, given the postulated time-scale (maybe 6 myr or about 400,000 gradually humanising generations) and the number of significant anatomical differences (as regards limbs, hands, feet, larynx, muscular refinements to lips and tongue, much larger skull, brain size and so on) could the necessary genetic revolution have conceivably occurred? Many changes need to be concerted. *In bacteria two or three coordinated mutations may occur but, even in the timescale of the universe, not enough to change the* **function** *of a protein let alone innovate physiological and psychological capacities (such as, in man, ability to speak, make music, conceive abstract principles or otherwise symbolically codify the world).* **It behoves an evolutionist to spell out which sequence, specific effect and quantity of single, 'beneficial', editorial-evading and germ-line mutations, piled positively upon each other and thus each rendering offspring significantly fitter than their parents, orchestrated the rapid 'invention' of a human.** Plain assumption or blank assertion that it *'must have happened'* cuts no mustard.

To put matters in perspective it is noted[112] that 150 years of bacteriology generating an optimal 48 generations per day would

---

[112]   see also *SAS* Chapters 21: Minimal Functionality and 23: Evolution in Action?

produce over 2,500,000 generations each populated by, at very least, millions of individuals. In human terms this number of generations would equate to more than 60 million years of history. In the simpler, prokaryotic case every type of mutagen has been systematically thrown at organisms and yet no major change of any kind has ever been observed. This applies to all other genetic experiments. Strains, races and, at maximum, new species are the most that's come about.

Thus, even remotely assuming *DNA* alone is capable of such transformation is it possible that this tally of *BM*s could, each in turn, be fixed in so few years? By what experiments has the hypothetical 'genomic leap', turning chimp program into human, been tested let alone validated? Or does the theory of evolution remain in an extremely speculative phase?

Computing experts, on the other hand, know how *top-down* programs are engineered, refined and, routine-wise, varied according to target function. And you never *start* with bugs though over time they may accrue. Is this how information-rich, heavily teleological and regulated code generates auto-reproducing bodies? From a *bottom-up* perspective it is not. Instead, 'beneficial mutations' to produce the 'right' ape > man transition must have worked 'specific magic'. These need, randomly appearing at an average speed of at very least 200 per generation, to become fixed without loss in each of a very long series of gradually-humanising populations. **It may well be argued that blind, natural forces do not now and never did achieve such high informative velocity.** Ape > man is over. Just as it's not happening now it didn't and it won't. **Thus, unless you counter the statistical assertion of required velocity with heavy speculation, human evolution is disproved.**

Hot-Spots and Humans.

If, however, the narrative just *must* be true what else is there to do? One ignores objections and research proceeds.

In this case, before interest turns to Y-chromosomal and mitochondrial *DNA*, let's consider seven more relevant autosomal issues.

Firstly, among those genes undergoing strongest chimp-human variation are those involved with immune defence. This is hardly surprising. *DNA* casinos have been established to perform sexual and immunological shuffles; these lend potential to adapt to challenges from the environment. But the dedicated lotteries that meiosis (with crossing over) and immune response employ do not *evolve* a thing.

Both are marvels worth consideration. For example, your internal police force, your immune system, plays its hand by generating 'not-you' combinations used as checks against illegal immigrants. If any immigrant checks positive against a 'not-you' permutation (as expressed by protein antibody) then the system swings into counteraction. Such complex shuffling mechanism, composed of several hundred 'variable domain genes' (blocks of immunological *DNA* code) and called 'somatic recombination', keeps your immune system on its toes. It works by recombining the blocks of code into new 'combination-lock' permutations. The recombination occurs during the maturation of B-lymphocytes and permits them to generate new kinds of antibody to fight new species of illegal immigrant, new threats of infection. At every level in this war the ability to recognise self from non-self involves code, signal, recognition and, above all, specificity. An immune system therefore consists of mechanisms that lock onto and variously disable pathogen-associated molecular patterns (often protein) and their accompanying body. It deploys a lottery that may generate up to a thousand billion different receptors to protect your health. *Such a system is, in overall construction, an information-rich, automatic and reasonable process, the very reverse of chance; its desired effect is deliberate variation which rapidly produces the recombinations from which a protein that confers resistance can arise.*

Human Leukogen Antigen (*HLA*)

If man was created, we do not know where. But if man was created with woman, an interesting fact has been pointed out by Francisco Ayala, a Professor of Genetics at the University of California.[113] He believes that perhaps 7 per cent of human genes show alleles.[114] In other words, this proportion of genes can show variations in the same character, for example, straight versus curly hair, brown versus blue or green eyes etc. On the basis of this 'heterozygosity' Professor Ayala calculated that the average human couple could have 10^{2017} children before they had one child identical to another. This number is far, far greater than the stars in the sky, the grains of sand by the sea, all organisms that have lived or atoms in the universe (about 10^{80}).

However, the likelihood is that Professor Ayala (a population and evolutionary geneticist) has greatly underestimated the number of alleles and effective genetic variations. *Top-down*, it would be conceptually rational to introduce maximum useful adaptive potential into any program written out as a genome. Such flexibility would involve initial sets of alleles (maximum four per particular gene, two in each sex), provident splicing regimes, regulated *RNA* editorial features and so on.

In the 1990's the Professor may have further muddied waters. He sought, by using choosing a mutation 'hot-spot' in the form of an immune

---

[113]   Ayala F. 'The Mechanisms of Evolution', *Scientific American,* vol. 239, no. 3, September 1978, p. 55.

[114]   see Glossary.

system gene and assumption-loaded calculations, to disprove a single pair of human parents, that is, to molecularly bust the ghosts of Adam and Eve.

An immune system antigen-binding site needs to recognize and bind many kinds of antigen (foreign substances inducing the production of antibodies). A protein called HLA-DRB1 has, consequently, many hundreds of alleles (corresponding to binding with different antigens). The hot-spot on this gene is restricted to part of what is called exon-2 (other parts of it are not 'hypervariable'). Professor Ayala, apparently ignoring the large possibility of somatic recombination at the 'lottery' of exon-2, treated the whole gene as normal and asked whether offspring of a single, original couple (whose maximum would be four alleles for each gene) could have generated such a profusion of alleles in the mythically short time since they are supposed to have existed.

Ayala proceeded to use data from the human, chimp and monkey (macaque) HLA-DRB1 gene to construct their phylogenetic 'tree of life'. Two implicit assumptions here are the 'fact' of common ancestry and that alleles caused by regular mutation would give a reliable timescale. He also calculated that 32 alleles were present at the time of chimp-human divergence (assumed ~6 mya); and that, to avoid deletion by natural selection or genetic drift, the average population would have to comprise ~100,000 with a minimum of ~40,000 individuals - far in excess of a single heterosexual pair.

This calculation was based on equations drawn from population genetics (such as the Hardy-Weinberg equation) dependent on simplifications, another 6 explicit assumptions (where any one being incorrect could skew the results) and a further implicit but incorrect assumption viz. the sole 'stochastic' cause of change is not an immune system 'lottery' but random mutations. Ignoring rapid immune system recombination guaranteed an overestimate of time required to generate the many alleles. For example, calculations involving *part of the same gene* called intron-2 gave a much slower rate of allele production and a minimal population of 7 (and average of 10,000).

A third, further study[115] using exon-2 and more intron segments radically revised Ayala's family tree. Particularly, exon-2 samples did not cluster into a species-specific bundle but the non-hypervariable intron samples did. Does this mean that exon-2 segments have been conserved for 30 myr when monkeys are assumed to have diverged from great apes? Or, as for example when chimp mates gorilla, there has been inter-species shuffling? Or, the most sensible meaning, that evolution is not the issue at all. Such non-species-specific exon-2s are

---

[115] Molecular Immunology 45 (2008) ps. 1250-1257.

exactly as you would expect if instead the issue is generating defence against the same antigens in different organisms.

Moreover, the associated regions on the chromosome of this gene do not appear to undergo meiotic recombination. Probably due to their necessary cooperation, specific alleles of HLA-DR (a dimer one of whose parts is HAL-DBR1) and associated HAL-DQ (another dimer) are inherited as a block. In humans there appear, overall, to be 5 of these 'blocks' (called haplotypes).[116] Of these we hold 3 in common with apes and monkeys, 2 are possibly from around the man-chimp divergence (~6 mya) but 1 is missing in chimps. **It is therefore possible, argues Anne Gauger,[117] that four haplotypes could have been carried, like alleles, on four homologous *DNA* strands of an original human couple.**[118]

There are those who swallow Ayala's argument whole but here it is pulverized. This example should serve as a statutory warning against, as in politics, taking any evolutionary interpretation of events or set of figures unchecked and stand-alone - even if they are warningly announced as 'unassailable fact', 'inevitable', 'beyond doubt' and so on. Too often this is simply empty threat.

Skin Colour

The third issue is skin colour. Dark pigmentation protects skin from the harmful influence of strong light (especially in the UV band). On the other hand, sunlight is necessary for the sufficient production of vitamin D. Clearly a balance needs to be drawn between exposure to strong sunlight and risk of vitamin D deficiency. If you believe that man was originally dark-skinned you must believe, in accordance with evolutionary theory, that mutations converging perhaps many times by chance have reduced skin coloration enough for fair skinned individuals to live, naturally selected, in (northern) areas of weaker sunlight and still produce sufficient vitamin. In other words, the complex, integrated and well-regulated metabolisms of melanin production, sun-tanning and vitamin D production occurred, Darwinian-wise, by chance.

Top-down, of course, the original human program would include sufficient adaptive potential for a flexible genetic response to climatic circumstance.[119] Not mutation but intrinsic, pre-programmed regulatory response is the key. This certainly involves more than ten genes, their alleles, associated supercoding mechanisms and the many substances involved in skin pigmentation (not least melanin). It must also involve

---

[116] see Glossary.

[117] Anne Gauger: Science and Human Origins Chapter 5 ps. 113-117.

[118] see also simple, footnote mathematics in chapter 'African Eve'.

[119] *SAS* Chapter 23: Neo-Darwinism Doesn't Work.

heritable interplay between these genes (such as 'dimmer-switch' regulation of output) and/ or heritable epigenetic setting.

No doubt, the variant nature of alleles is more clearly understood than the programmed subtleties of the 'dimmer-switch' regulation of gene expression. *What is also clear is the great pre-arranged adaptive potential afforded by such regulation as regards fine-tuned, flexible response to circumstance by transcription and translation splicing systems, micro-RNA and transposon switches, epigenetic markers and other programmed subroutines.*

Skin and hair pigmentation are one thing but only a single 'mutation' on a single gene seems sufficient to separate thick-haired eastern from thin-haired western Eurasians. Why should such variation be, rather than intrinsically programmed, presumptively 'mutant'?

Of course, for those who reject the idea of the informative projection of a program (that software buffs all understand) only an accumulation of accidental mutations, constrained in their survival by natural selection, can explain the origin of programs. It is accordingly plausible that an evolutionary interpretation of adaptation by mutation or comparisons of homologous *DNA* or protein might seem to indicate the way programs are connected by inheritance through different creature forms. By such circular reasoning you might point out genetic correlations between chimps and humans that 'proved' their common ancestry. Perhaps even more persuasive might seem useless but identical strips of 'garble' found locked like 'prehistoric flotsam' in the database of both types. *If these really consist of random mutations and amount to identical bugs it could indicate that apes evolved to mutant apes.* After all, who in their right mind programs garble in among their lines of working code? So ancient bugs passed down through generations might easily compel belief in Darwin's theory.

In short, nobody argues that bugs can't develop to disorganise a program; degradation from an orderly initiation will occur. Just as physics deals in orderly beginning from a transcendent projection, so does the bio-logic of design.

On the other hand, for the theory of evolution a reverse form of logic applies. Evolution argues random interference in a working message may *improve* it; bugs *build* systems. However, as well as mutations that 'drive life's progress' other neutral, useless ones are also passed along. The latter might be called 'useless junk'. **So, to some, the presence of species-common 'junk DNA' might seem an 'indicator', even 'proof', of evolutionary lineage. Is this the case?**

Non-protein-coding (ex-junk) *DNA*

The genome is a vehicle of information. Every vehicle takes, of course, some wear and tear and so some damage that is rightly called 'junk' can be expected; but how much can its healthy operation bear? Who cares? In an evolutionary paradigm you welcome accidents. Accidents (mutations) are what build vehicles; they are evolution's microscopic engine. Some might appear to confer advantage; others might not but still, perhaps, provide a careless duplicate or unused space on which further accidents might somehow accumulate to the point of innovating something. This, though central to evolutionary theory, is pure guesswork.

Non-protein-coding *DNA*[120] used to be called '^jnu~%n*k'. Now it is established that, although only about 2% of the human genome codes for functional proteins, the vast majority of the rest is transcribed as *RNA* elements or used in structural capacities. Indeed, so intensively 'alive' is the DNA computer that[121] 'evidence indicates that most of both strands of the human genome might be transcribed, implying extensive overlap of transcriptional units and regulatory elements'. *This is all nothing to do with evolution and all about function.*

Since bananas as well as humans respire, reproduce and develop you might expect they and other eukaryotes had many genomic features in common which also have nothing whatsoever to do with evolution and everything with efficient operation. You may, therefore, find it less than curious that some sections of ex-junk are, for whatever reason, practically identical in humans, elephants, dogs and marsupial wallabies; indeed, nearly 500 sections of over 200 bases have already been found identical in mice, rats and humans. There exists, moreover, a strong correlation between the placement, though not the actual sequences, of short and long interspersed nuclear elements (*SINE*s and *LINE*s)[122] in rats and mice. Perhaps mammals each have their own repertoire of once-called-useless chunks of *DNA*.

Chromatin, far from junk, is a most important combination of the *DNA*, *RNA* and proteins that compose a chromosome; and a chromosome may be seen as one volume, accessed in an automated way, of the encyclopaedic genome. Chromatin is closely bound up with the regulation of gene expression. Any effect on its organisation impacts on this life-critical operation; and it is known that n-p-c (previously 'junk') *RNA*s help compose such organisation and can affect gene expression by modifying it. For example, various proteins and n-p-c *RNA*s mediate the process of correctly attaching a vital transcription complex, *RNA* polymerase, to *DNA*.

---

[120] see *SAS* Chapter 23: Non-protein-coding *DNA*.

[121] 'Genome-wide transcription and the implications for genomic organisation': Nature Reviews Genetics 8: 6-2008 ps. 413-23. Also Science 319: 28-3-2008 ps. 1787-89.

[122] see Glossary: transposon.

Banding patterns on mammalian chromosomes reflect systematic densities in a compartmentalised way that resembles bar-coding. So-called 'white', loosely-packed R-bands (called euchromatin) are variously rich in the genetic letters G and C, occur where there is a high concentration of *SINE* elements, active transcription of protein-coding genes and replication early in the process of cell division. On the other hand, 'dark', tightly-packed G-bands (called heterochromatin) are rich in the letters A and T, transcriptionally inactive, occur where there is a high density of *LINE* elements and replicate late. In short, repetitive elements may represent a strategy for controlling chromatin.

In that case are AREs (ancient repetitive elements), as some suggest, coincidental accumulations of repetitive junk *DNA* passed like ineffective rubbish down an evolutionary chain? Did n-p-c repetitive *SINE*s and *LINE*s coincidentally accumulate about the same but also differently in each abovementioned rodent's banding case? Perhaps, it is argued, these tracts of uselessness are just analogous to ancient data traces of which are still left on a hard-drive after the delete button has been hit. As hard-drive may be copied in entirety, so is genome replicated. You could float ancestral rubbish indiscriminately down an evolutionary stream. *Or, on the other hand, has some higher order pattern been passed down rats, mice and other mammals, although each of the elements is systematically but differently coded, from a common ancestor?* It is unlikely 'crazy' chance could work in such an ordered way. You might better presume that it has resisted natural selection, been conserved and therefore has some purpose. *Very often conservation is evidence for function. It suggests a necessary use.* In AREs and chromatin you definitely have it.

Repetitious tandem and interspersed n-p-c *DNA* occupies up to 50% of the human genome. Further jobs already identified include X chromosome deactivation, constitution of a histone-positioning code, repair of broken *DNA* and involvement in placental development. Transposable elements are also involved in cell stress responses, chromatin condensation and *DNA* methylation required for epigenetic super-coding. Some uses even appear sequence-independent, such as stretches of 'tandem satellite' repetitive *DNA* sequences that serve to compose centromeres (critically linking chromatids in cell division and sexual recombination) and chromosomal caps called telomeres. Such 'junk' also repairs *DNA*, assists replication, regulates transcription, controls editing and splicing and generally performs the sort of operations you might expect to occur in a high-grade, circumstantially-flexible computer program.

Can a Darwinist be wrong - ever? By 2009 Richard Dawkins was still observing (The Greatest Show on Earth ps. 332-3) that 95% 'junk' *DNA* was an embarrassment for 'creationists'.

The ENCODE project (**ENC**yclopedia **Of DNA E**lements) is a public research consortium launched by the US National Human Genome Research Institute after completion of the Human Genome Project to find all functional elements in the human genome. In 2012 a Cambridge co-ordinator, Ewan Birney, was quoted as saying that the genome is 'alive with switches, millions of places that determine whether a gene is switched on or off'.

In September 2012, with ENCODE's announcement, Dawkins' tune changed. Now 'junk' was no longer evolutionary jetsam and flotsam but, as critical and refined coding, 'just what a Darwinist would expect'!! A twist, a U-turn flipped without so much as the wink of a cheeky eye!

In fact, we know that non-protein-coding regions of *DNA* engage in multiple functions. We can add to our list that they generate tens of thousands of short n-p-c *RNA* features (such as micro-, pi- and si-*RNA*s) that seem involved in the genetic operating system and thus regulate the addressing, transcription, translation and post-translation modification systems by which genes are expressed as protein. The non-junk answer therefore seems to be, in a word, control; regulation of expression; a system operating so that exactly the right protein is produced in the right quantities at the right times in the right cells. Such tightly controlled expression is species specific and, within species, cell specific. This is no small order in a space far, far smaller than the eye can see.

Organisms transcribe much of their DNA, including so-called junk, into RNA. Over 90% of bases from small, sample sections were, by 2007, found to transcribe by coding either for protein or *RNA*. While perhaps 22,000 human genes that code for protein have been revealed there may be up to 450,000 '*RNA*-genes'. What's the point of them? *Why waste energy transcribing them if they amount to informatively useless and thus interfering junk?* Francis Collins, Leader of The Human Genome Project, noted some time ago that only a small fraction of such factors are known to be useful but his appeal to evolutionary 'junk' status for the rest simply assumed unproven non-functionality. Discoveries have well overtaken this opinion. It is now defunct. Definitively, 'junk' is junk.[123] **Already at least 80% of the genome is already judged biochemically functional. There is method very far from randomness, very far from mutant madness in the integrated patterns of a cell's complexity.** *In no way does 'junk' DNA support the notion of human (or any other) evolution.*

Pseudogenes

Here's another kind of 'junk'. So-called pseudogenes are part of the evolutionary 'junk' mindset. Each was thus presumed to be the genetic

---

[123] Nature Vol. 489: 6-9-2012 ps. 57-74.

corpse of a once-functioning gene. It was argued that a working gene may have been functionally maimed, disabled or killed by such mutation 'broke' a promoter sequence or an intron; or perhaps it caused a replicatory brake-signal to fail, or faulty reinsertion of m-*RNA* by a retrotransposon such as may produce what is identified theoretically (using sequence data) but less surely by experiment as biologically non-functional material. However, are all pseudogenes 'broken genes' or do some at least have useful function?

Maybe presumed but theoretical malfunction has led to imperfectly computerised methods of identification because now many have been diagnosed functional. And functionality would, of course, render the evolutionary interpretation of such pseudogenes, inherited as junk, irrelevant. In fact, as well as finding that a large proportion of non-protein-coding *DNA* (maybe up to 90% of *both* strands) is transcribed to *RNA*, the ENCODE project has revived many pseudogenic corpses. Working pseudogenes produce useful protein, express non-coding *RNA*, help in gene regulation and act as tissue-specific switches, silencers or 'scaffold' proteins needed in development. Indeed, bio-informatic analysis has generated a genome-wide survey of biologically functional pseudogenes.[124] It is likely that the number genetic resurrections will increase.

Maybe non-protein coding function is sometimes transient or contextual but some 'junk' produces viable protein; and, if not protein, functional *RNA*. Other pseudogenes may not produce protein but important *RNA* factors; they may generate signalling agents but even without using *RNA* at all influence chromatin dynamics, help regulate recombination and mediate in gene splicing. *They are what a programmer would identify as essential switching, silencing, modulating, sequence recognition and feedback elements.* All this deletes the pseudo out of pseudogene!

Gene conservation means that certain genes and sequences are preserved intact in different organism. This most likely indicates important function. Some pseudogenes are extremely well conserved, that is, appear the same in different species. Either this highly conserved kind of waste is so inert that it endures below natural selection's radar; or it isn't waste at all. For example, a protein that is part of a tumour suppression system is transcribed from a pseudogene. Indeed, some functional sense has already been made of 7000 such 'vestigials' and 10000 long n-c RNAs (Nature 465: 24-6-2010 ps. 1016 and 1033). And thousands of mouse 'pseudogenes' have been found active. In this case, there is no need at all to interpret a pseudogene

---

[124] Svensson O, Arvestad L, Lagergren J: PLoS Comput. Biol. **2** (5-2006).

conserved in mammals as far removed as mice from men as evidence of common ancestry. Why, after all, should so-called 'mistakes' persist - unless they are neutral in effect *or* not random and conserved because they've definite jobs to do? In other words, mutation's entropy of information may take some toll but why write off pseudogenes as 'junk'? **They would simply, like other forms of 'junk *DNA*', need to be perceived in a new light.**

Suppose a gene was once operative and its 'corpse' is a product of miscopying or other genetic wear and tear; and, rather than being selected out as useless, its inert neutrality has lain passively inherited through evolutionary transformations. However, why, for example, does a single 'pseudogene' (out of four genes needed) stymy the production of vitamin C in some types of monkey, bat, bird and fish; also chimps, gorillas, guinea pigs (but not rats) and humans (except at the foetal stage)? If the issue is regulatory (as possible synthesis of vitamin C by infants might suggest) then perhaps some humans are still able to synthesize the vitamin. Otherwise, although no function is known for this conserved pseudogene it may turn out to have one. Even if it does not, deactivating mutations may or may not have occurred by chance in the same weak 'mutation hotspot'.

Another pseudogene, the so-called β-globin pseudogene, also shows great 'conservation' or 'stabilisation' of structure in mouse, cow, rabbit, chicken and all major primate groups including man. The fact that 'breakage' occurs at the same place in chimps and humans is often interpreted, in evolutionary style, as meaning that a functioning gene or extra copy of it mutated in a male/ female pair of imaginary ancestors; and that no non-mutants survived. Since natural selection should have cleaned it up, what advantage does this much-conserved 'genetic trash' convey? We've noted conservation is interpreted to mean a gene may serve some important role; and congruence of genotype or phenotype may just as well result from common blueprint as descent. Certainly neither we nor chimps appear to suffer from an apparent 'silence' of the β-globin pseudogene. Therefore, given wholesale but ignorant 'junking' of n-p-c *DNA* in the 1980's, why compound error and dismiss 'pseudo-sequences' whose meaning isn't perhaps as yet perceived? Is it transcribed to *RNA*? Does it mediate as a switch between production of foetal and adult haemoglobin?

In short, why have some genes been truncated, rendered apparently useless but conserved at the same position in mice and chimps and men? Would vitamin C conservation-by-accident across eons of macroevolution show that, far from a rat, you had descended from grandparent mice? Or are pseudogenes (and repetitive elements) factors whose uses we do not yet properly understand? Clearly you could interpret the presence of 'apparently ancient' relics at the same hot-spots on mouse and human genomes either as a case of evolutionary lineage *or* of critical function. It is only by *assuming* the former, that is, non-functionality, that any description of so-called pseudogenes as genetic graveyards is maintained. However, as The Computer Analogy has suggested, perhaps most are a facet of the genome's operating system - one whose super-coding finely regulates expression of the genes.

Fusion of Chromosome 2

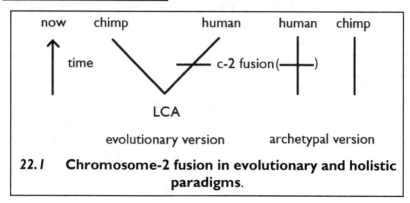

22./   Chromosome-2 fusion in evolutionary and holistic paradigms.

It is not the number of pages or volumes in a *corpus* that counts; it is the type and quality of information it contains. Thus you might understand the complete jumble of chromosome numbers. There is no evolutionary sense to be made in any kingdom. For example, chimpanzees, gorillas and orang-utans have (along with wild tobacco, potatoes and the pretty corncockle flower (*Agrostemna*) a count of 48 chromosomes; you have 46, as does the sable antelope and an arctic clubmoss called *Diphasium complanatum*. A Rhesus monkey has only 42, a horse 64, carp 104, pigeon 80, hedgehog 88, green tiger prawn from the Persian Gulf 90 and a freshwater shrimp (*Atyaephyra desmarestii*) 32, earthworm 36, yeast 32, rat 42, aquatic rat 92, horsetails 216 and the adder's tongue fern over 1200! No evolutionary numerical connections are discovered in plants, animals, fungi or protozoa. What did you expect? Ask any academic which is more potent, information-wise. Is it a slim volume of concentrated knowledge succinctly and coherently expressed or thousands of pages of ill-defined waffle? *Quality of information and not quantity of DNA is, as the Dialectic certainly confirms, the real criterion by which to judge a manual's source.*

However this may be, ape chromosome count is 48 and human 46. Some think this difference evolved because fusion of so-called elements 2A and 2B (chimp chromosomes 12 and 13) are represented by a single chromosome 2 in man; and that genetic inspection of such fusion 'proves' that man evolved from ape.

Why? Whether or not humans evolved, such a fusion must have occurred *after* LCA, last common ancestor or man-chimp split. Therefore, the human complement of 46 might have started as 48 and be the product of post-divergence fusion or, if chromosome 2 was always so, might not. *What has evolution got to do with it?*

Nevertheless, if you presume that ape > man evolution must have happened, how might your detailed (but, by now, less relevant) analysis proceed to 'prove' it? Is its *bottom-up* construction right? If you're still interested then hang on, if not skip to section end.

Firstly, an outdated assumption is made that so-called 'junk *DNA*' is useless, mutant rubbish from an ancient evolutionary path. This has been shown by later science (such as the ENCODE project) patently untrue.[125] Although residual mutations may endure the great majority of so-called 'junk' plays a critical role in program-switching, that is, the regulation of genetic expression. The assumption is redundant.

Secondly, we discuss the apparent fusion of chromosome 2.

What, in a chromosome, is a *centromere*? It is a spindle attachment, that is, the region at which a chromosome attaches to the spindle during the mitotic process called cell division. Chromosome 2 seems to show centromeric traces at the two points expected if chromosomes 2A and 2B (each with its own centromere) had joined; but one has been disabled because, in mitosis and meiosis, only one centromeric junction is required.

Next, what is a *telomere*? It may be a repetitive *DNA* sequence (in humans and apes TTAGGG) that 'laces' the ends of linear chromosomes so that they do not 'fray'; and it may also be a marker that defines an organism's natural span of life. What could be telomeric elements have been found in chromosome 2 where the junction of chromosomes 2A and 2B might have left them.

Finally, recall that each nucleus contains 2 copies of each chromosomal type, one from each parent.

Change in chromosome number is, as with Down's syndrome, mostly damaging and rare; thus, to become fixed in the entire human population it would have had to be, besides rare, at least neutral and probably advantageous. But the supposed mechanism of fusion, head-to-head telomeric overlap and treatment by a recombination enzyme as if 2A and 2B were one broken chromosome, does not seem to be occurring with other ape or human chromosomes. And the fact is that 2-to-1 conjunction is not anywhere at present pushing apes from trees onto the path towards man. No ape-men keep cropping up by fusion's accident. This is not because the process cannot happen (the Indian muntjac, for example, has

---

[125] *SAS* Chapter 23: Non-protein-coding *DNA*.

6 chromosomes where other members of its genus have 48). It happens in many creatures. And if the first few humans were of count 48 it may have happened very early in our prehistory. In this respect the answer, fusion, may be right but, if at all, *after* any split from chimps.

However, those who promote such explanation may omit to mention just how chancy the event must be. Firstly, all information needs to be retained. Secondly, meiosis would be interfered with since there would exist only one version of 2, 2A and 2b - a haploid and not the normal diploid quantity. How, for example, could bivalents line-up on the spindle? Therefore, the unique mother sex cell of a gonad inside which any fusion must have happened may or may not yield a viable gamete that includes the new, fused chromosome. Unviable, infertile alternatives are more likely but one luckily holding 2 without either 2A or 2B would obviate the need to later lose the latter, now extraneous, pair from our genome. This loss is not informatively necessary since the information in them has been duplicated; but if the fusion event occurred they have been lost from all human genomes without exception. How otherwise might such a total loss occur?

Secondly, a viable gamete from this sex cell would have to be the very one which, unless its fused load were lost, was straightway fertilised to create a zygote. Thus an egg with accidentally fused chromosome would have a single chance of fertilisation in its female owner's life-time; if was a sperm the chance would be one in myriad billions. Thus the chances of losing the fused chromosome 2 prior to any offspring are very high - but never let improbability restrain a guess!

If the 'fusion gamete' held a copy of 2 alone then, following syngamy with a normal 2A-2B individual its chromosome count would be 47; and in the case of 2 with 2A *or* 2B it would be 46. However, the meiosis of such an organism would again be interfered with and it would most likely be sterile. At least, transmission could possibly involve the same cycle as the original 'fusion gamete', that is, with a gamete containing normal 2A and 2B from the other sex. Only if the mating was incestuous between fusion-containing parent and immediate offspring or between a brother and sister could a diploid case of chromosome 2 occur. In other words, incest would have borne our race because if this pair had viably mated twice or more they could have possibly created primal humanoids (chromosomes 2 + 2, like us); they could have theoretically created disparate human youth, both Adam and Eve, in a single generation. You'd suddenly have primal humans in amongst a troupe of apes; wild children would be educated in a truly feral way. Non-sterility is possible, incest is possible but the chance of all this happening is, by this stage, very, very slim. And, if there's any trace of 2A or 2B left, they will still have to be eliminated wholly from the pre-human germ line.

If you doubt it do the experiment! You can't, by law, extract chromosome 2 and insert chimp 2A and 2B into a human egg or sperm - but you could do the reverse to a chimp. Would half-humans appear, born in the cage - which you might mate to find, in little more than a decade in the same cage, a real and erect sprinkling of Adams and Eves? Or, due to morphogenetic constraints and genetic difference (especially in n-p-c *DNA*), are millions of additional changes necessary? The absurdly extreme improbability of this number of random but at the same time advantageous mutations occurring in 6 million years was discussed in the section 'Autosomal *DNA*'. However, in both chimp and man chromosomes 6, 13, 19, 21, 22 and X show the same banding while significant 'non-coding' regions of difference exist. The Y chromosome differs considerably. There are also segment inversions in chromosomes 1, 4, 5, 9, 12, 15, 16, 17 and 18. In chromosomes 3, 11, 14, 15, 18, 20 content may *look* the same; and the rest, except 4 and 17, are also somewhat similar. *These discrepancies aside, why should a joined or separate chromosomal situation make any difference if full genetic sequences are present in each case?*

Of course, to avoid this raft of improbabilities you might cut to the chase and speculate that the same fusion occurred simultaneously and identically by twin strokes of chance in the mother sex cell of a male and female who timely mated. In this mating the mutant gametes chanced to be the very two that fused and, thus fertilised, developed with the normal, diploid state of chromosome 2 - except that the remaining diploid chromosomes 2A and 2B would still have to be lost. In other words, did the unusual fusion process, always involving only chromosomes 2A and 2B fusing at exactly the same genetic location, occur just once in a single individual? Simultaneously in a couple of human precursors 'Adam and Eve'? Or, involving 'Adams and Eves', did it happen multiply and separately both in time and geographical location? Which of these far-fetched, magic-moment scenarios seems less miraculous, that is, which one more plausibly perchance supports the evolutionary mandate that such chromosomal fusion must have happened?

Must it? A programmer may need certain sub-routines but what prevents combining tracts of code? In this respect it also needs to be considered that the composition of ape telomeric *DNA* is restricted, that is, species-specific and containing different quantities of satellite *DNA*. Large amounts of such *DNA*, the hallmark of every other instance of fusion, is found on the tips of 2A and 2B but has, at the assumed point of fusion in 2, 'gone almost awol'. It is, for some reason or for chance-based none, reduced and degraded. Moreover interstitial 'telomeric *DNA* repetitions' are found throughout mammalian genes but have nothing to do with fusion - except by interpretation in this particular, 'evolution-proving' case.

Moreover the very location of fusion has been identified as part of (or at least within) a functional transcription factor binding site. Such sites help regulate gene expression. This instance is on an important, actively transcribing gene, the *RNA* helicase regulator called DDX11L2. This gene's transcripts are variable, complex, contain a number of mi-*RNA* regulatory binding sites and are associated with a variety of cellular processes. It operates in 16 out of 31 major tissue types including reproductive tissue. How, in this case, could head-to-head fusion have happened here? And how did such 'lucky' accident at the same time create (or at least not destroy) an important functional gene? *Yet this or similar unlikelihood, along with millions of other mutations, would be needed for the evolution of humanity. Such improbable, microscopic randomness to wield such vast effect! If not Star then thank your lucky stars you're here!*

Eighteen copies of the important DDX11L gene exist in humans but only two in chimps and four in gorillas. Moreover, genes and putative pseudogenes surrounding the fusion site do not exist at the ends of chimp 2A or 2B material. Also, a size discrepancy of about 10% (24 million bases) separates the smaller chromosome 2 from the sum of 2A and 2B. Why? And at least part of a large section of chimp 2B sequence information that has disappeared from human chromosome 2 is not lost but found, homologously, on chromosome 10. *All these observations weigh cumulatively against the process of fusion ever having happened.* If you delete it you are simply left with human chromosome 2 intact from the start.

Human Accelerated Regions (*HARs*)

Chimp and human programs are, as you'd expect, replete with similarities but containing salient distinctions. For example, human accelerated regions (*HARs*) are segments of the vertebrate genome highly conserved but remarkably distinct in humans. Of these *HAR1*, important in the construction of cerebral cortex, is a group of 118 bases. Within this group humans host an accumulation of so-interpreted 'mutations' that outnumber the differences found between other very different creatures (e.g. bird and bonobo) by almost ten times. Similar divergence occurs in other genes involved in the construction of brain (arguably man's greatest difference from all other organisms). **Why, though, should this have anything to do with evolution?**

Interestingly, this rash of changes does mostly involve regulatory elements rather than straightforward protein-coding genes. So also with *HAR2*, an intron on chromosome 2 that involves the foetal development of manipulative hand and wrist bones (yielding critically distinct human dexterity) and locomotive ankle and foot bones (involving gait and bipedalism). In these cases so-called 'mutations' affect the method of construction rather than materials. They direct the algorithm. *Top-down, the basis of biology is information. Natural Dialectic would not, therefore, suppose 'mutation' at all, rather appropriate refinement of archetypal subroutines with switching mechanisms that will make a man not ape.* **Why (except through a theoretical skew of mind-set) interpret such extra or different regulatory switches as 'mutations'?**

The recruitment of purpose, not random mutation, is intrinsic to such a view. Origin depends on the way - archetypal program or chance disturbance - in which you look at natural history.

The Male Chromosome: Y-*DNA*

The male Y-chromosome (called, as is the female X, a heterosome or allosome) comprises ~60 million base pairs of which 5% are in common with the X chromosome. Only this 5% can cross over in meiosis and are thus subject to recombination. The 95% male specific region (MSY), which includes 78 protein-coding genes, cannot recombine with *DNA* from its X complement. This means its sequence is passed unchanged from father to son. It remains, in the patrilinear manner of a male surname passed down in marriage, stable as regards ancestry. No lottery. Just a direct, unbroken male-only line of inheritance from Y-chromosomal Adam (from whom all men have stemmed) to the present time. Changes only occur by accidental mutation. Thus, clear of recombination's incalculable blur, such patrilinear descent nicely indicates, by comparison of mutations alone, not only the amount divergence of one Y-chromosome from another but also its owner's possible geographical location and (if you apply a molecular clock) putative date. Thus, although only about 2% of a man's total *DNA* it can act as a powerful marker in defining lineage.

If the chimp-human split from a common ancestor occurred (as is currently believed) about 6 mya it would be of interest to know the genetic sequence of this missing link; also, sequences for *Australopithecines* and *H. erectus*. We actually have samples for a few Neanderthal, Cro-Magnon and archaic *homines* (e.g. Denisovans who, interestingly, are linked by *DNA* signature to present-day Papuans). And we have, of course, sequenced current primate species including chimps and ourselves.

Is chimp Y-*DNA* close to ours? Although, according to evolutionary imperative, pre-chimp and pre-human lines diverged about 6 mya a definitive study[126] has revealed that chimp and human Y-chromosomes 'differ radically' in both structure and gene content (male chimps have

---

[126]   Nature 463; Jan 2010 ps. 536-539.

less genes than men). Such 'remarkable divergence' means the two sorts of Y are as different from each other as human from chicken chromosomes! The specious explanation given is that, in chimp and/ or man, for no particular reason this tiny percentage of the genome alone embarked on an evolution spree, that is, exceptionally rapid mutation as opposed to relative conservation in the rest of the genome. Of course, hypotheses abound to try and counter such confusion but never currently, on pain of a fall from High Table, include the obvious anathema of difference by original design!

Nevertheless, the Y-chromosome can, using certain assumptions, be used as a male-line 'molecular clock'. Like the female-line mt-*DNA* we'll shortly meet, it can be used as a space-time tracker of the species *Homo sapiens*. The clock is based on ticks defined by a single base mutation (an SNP) in a male or female haplogroup (see Glossary). Persons with the same mutation are classed in a 'haplotype'. A group of haplotypes (a haplogroup) is defined by a 'marker' (a *DNA* sequence of known location) containing a specific SNP mutation or mutations. It is assumed this SNP was held by (using mitochondrial *DNA*) a matrilineal or (using Y-*DNA*) a patrilineal common ancestor. In this way a theoretical, intra-species phylogenetic tree can be built and related to current geography (e.g. Japanese men are in one haplogroup and Congo pygmies another), presumed antiquity (the rule of thumb is 'more diverse, older') and possible, historical migration routes.

Of course, with such a rough and ready method both speculation and disagreements abound. The discovery of more mutations, previously unknown haplotypes, undiscovered haplogroup divergences, mistaken assumptions (e.g. dependence on an SNP mutation that could happen more than once) and/ or revision of the clock's tick (mutation rate) could all force radical changes in interpretation of the data. In other words, it all depends on which markers you choose and the molecular clock speed you adopt.

Nevertheless, the current consensus is that the address of Y-chromosomal Adam or Y-MRCA (Adam's surname meaning Most Recent Common Ancestor) was not paradise or even Eden Garden. He roamed, as the first anatomically modern man, in Africa. This is guessed because greater genetic diversity is found here than elsewhere in the world. The basal divergence is thought to be between bush-men and the rest of us including all other Africans.

Again by current consensus Adam's time is calculated at between 200 and 300 kya. However, haplogroup discoveries keep shifting time so that other research has assigned ages between 400 and 40 kya.[127]

---

[127] Whitfield *et alii*; Nature 378: 23-11-95 ps. 379-80.

Such veering guesstimates attracted a median age of about 210 kyr but this doubtless will, with new findings and methods of calculation, continue to oscillate.

Another study[128] using a larger sample but a much shorter *DNA* sequence, found remarkably little divergence within any primate species; at the same time, paradoxically, it found significant divergence between species. Internal lack of divergence would indicate recent origin of the species; external divergence would indicate an ancient common ancestor or separate, archetypal creation.

Mt-*DNA*

Ancillary *DNA* is found in single-ring form in the mitochondrial organelles of cells. Mammal mt-*DNA* is much smaller than Y-chromosomal (about 16500 base pairs representing ~0.00006% of total genome) but greater in quantity. Human cells contain between zero mitochondria (red blood cells) to more than 1000 (liver cells). A fertilised egg has one diploid nucleus but 100,000 mitochondria (a sperm has about 1000); thus mitochondrial *DNA* is high in copy number, ~150,000 copies per oocyte but only ~50 in sperm.[129] Also, lacking its own sophisticated editorial and repair mechanisms, an mt-*DNA* sequence is more prone to mutation than 'spaced out' (more slowly mutating) nuclear Y-chromosomal *DNA*. Therefore, because the 'evolution rate' is relatively rapid its larger number of mt-*DNA* mutation points makes tracking easier and more refined. Of special interest is the fact that, as opposed to Y-*DNA* tracking patrilinear ancestry, mt-*DNA* is used to trace female lines (that is, as if a matriarchal surname was passed down in marriage). The reason for this is an assumption that, in the process of fertilisation, sperm mid-pieces and tails (and thus mitochondria) are destroyed. Only the female's organelles survive and thus all offspring, male and female, possess only mother's undiluted mitochondrial *DNA*.[130]

The hope is therefore, to use this way of better understanding the 'phylo-geographical' (geo-historical migrations and isolations) origins of various populations and races of *Homo*; and tracing lineage to a clear 'mitochondrial Eve', first human mother of us all. *Thus, along with its other useful features, mt-DNA's strict matrilinearity is central to the 'African Eve' hypothesis.* This we turn to.

---

[128] Nature 385; Jan. 1997 ps. 125-6.

[129] how can it be explained that different tissues are informed to produce necessary, precisely relevant variations of mitochiondrial protein?

[130] see Chapter 23 for further discussion of this dogma.

23. African Eve

Does age mean beauty? Could antique ladies steal the show? One definitely fires passions - so-called 'Mitochondrial Eve'.

In 1987 **'African Eve Theory'** was rolled, with customary fanfare, off the blocks. All propaganda presupposes that repeat publicity will generate its own momentum. Add to this the authority of science and you have what is perhaps the currently favoured explanation for the reason why you're here. Eve - a pressing substitute for Our Lady of Eden - resolves the issue in a quite semitic way. Time-scale apart, the idea claims all humans rose from just one 'lucky lady' (or a population of her peers). On what evidence is theory based? Did Mum actually come from Africa?

The theoretical model was developed by Mark Stoneking, Rebecca Cann and Allan Wilson at UC Berkeley, USA.[131] Much work has been done internationally sequencing contemporary mt-*DNA* and the latter couple dated 'Our Lucky Lady' at 200 kya by the use of an mt-*DNA* 'molecular clock'. Also in 1987 an initial and impressive global mt-*DNA* tree was published to show how, by percentage sequence and apparent linearity, populations on the various continents seem to be related. How this has come to be is not so clear but, by postulating Eve's haplogroup as one called L0 it seems that Africa, also sporting haplogroups L1, L2 and L3, hosts the most diverse (and, therefore, hypothetically the oldest) species of mt-*DNA*. No indigenous groups outside Africa (M, N and their sub-groups) have L-type; none in Africa, except in the north-east around

---

[131] Nature 325: 1-1-87 ps. 31-36. The theory is also called Recent African Origin (RAO) of modern humans, Recent Single-origin Hypothesis (RSOH), 'Replacement Theory', 'Mitochondrial Eve' and 'Out of Africa'.

Egypt, have M and N. It is therefore supposed that M and N were derived from a hypothetical expansion of the population of L3 carriers (cause unknown) at the same time as a theory-dictated exodus 'Out of Africa'.

This, given the racist undertone of Darwinian 'inferiority' and 'domination by the fittest' allied with the social evils of slavery and political tyranny, appears a satisfactory, politically correct solution - our story is as unifying (almost) as creationism's one. We are all at root from Africa; and, as mt-*DNA* scans seemingly imply, born from a 'population bottleneck' of one or a few Mums.

Is this, a fashionable view, correct?[132] Maybe, maybe not but the ways in which its speculation is upheld need investigating. Before engaging such investigation let us simply note that it is difficult to predict phenotype from genotype; either ancient or modern, *DNA* alone cannot exactly define its owner's body-shape. Secondly, teenage girl 'Denny' is supposed to represent hybridisation between a Denisovan father and Neanderthal mother (as if you might suppose a child of African and Caucasian parentas 'hybrid' too?); the knowledge that Neanderthals, Denisovans, the Dmanisi crew, *H. sapiens* (ourselves) and possibly other as yet undiscovered strains of *hominin* interbred means that current conjecture finds human descent more *clustered*, less divided than the evolutionary imperative of splintered, tree-like speciation requires. Thirdly, if humans ever sallied forth from Africa they must have sallied back again. Is this mix, *pace* theory, indigenous or due, as theory requires, to perhaps a more recent back-flow? The Khoisan people of Africa's deep south are thought to have split from other humans well before Eve's northern exodus yet, carrying the L0 haplogroup (Glossary), also be an ancient source of human lineage. However, Eurasian genetic presence has been found in these genetically isolated, click-language Khoisans; also a Neanderthal trace in the west African Yoruba population.[133]

So, was 'back-flow' or 'no-flow' at all the case? Was Eurasian and Neanderthal programming global, that is, perhaps archetypal? Or are we compelled, for Darwin's sake, to invent stories of recent, reverse migration (that is, immigration) from the north? If you seek muddle look no further than your human anthropology. We can build stories like a house of cards.

Mt-*DNA* sequences have been obtained from humans worldwide, chimpanzees and other apes; also from Italian Cro-Magnon, 'archaic' Australian, Siberian Denisovan and Neanderthal bones (the latter from German Feldhofer, Croatian Vindija and Spanish Sidrón and Atapuerca caves). But how reliable is speculation born of mt-*DNA*? Does it

---

[132] not forgetting European *Graecopithecus freybergi* (Chapter 20).
[133] Research at Cambridge 8-10-15: 'Ancient genome from Africa first sequenced'.[133]

reliably identify the date Our Evolutionary Mother must have lived (perhaps between 200 and 130 kya)?

Why, it may first be asked, should only a single Eve's line survive? Perhaps because there *was* only a single Eve - unless we guess at several sallies out of Africa (between, say, 200 and 65 kya). In which case, have all except one of such multiple, evolutionary Eves lacked an unbroken chain of fertile, female offspring? Astronomically unlikely but, unarguably, possible.

Next, mt-*DNA* is susceptible to both mutation *and* (although it is supposed to pass exclusively down the female line) some level of sexual recombination.[134] Adam's presence would distort mitochondrial time-keeping[135] and tracing origins through a single Eve-tree be impossible. Indeed, such male leakage into a female zone would cast doubt on all the research that uses mt-*DNA* as a molecular clock. Therefore, a key necessity for 'African Eve' theory is that strict, non-Mendelian matrilinear mt-*DNA* pedigree is maintained.

However, could paternal 'leakage' happen? Of the 1-2% proportion of spermatic to zygotic mitochondria could none slip through? Even a cumulative fraction of a percent per generation would render theory opaque. Actually,[136] John Maynard Smith and others were systematically ignored, to his frustration, by an African-Eve prone establishment, for suggesting that paternal and maternal mt-DNA could be swapped (in chimps as well as humans). Moreover, lack of paternal mt-DNA destruction has been discovered in sheep and humans. In 2002 (Schwartz and Vissing[137]) a large proportion of paternal mt-DNA was found in a patient's muscle tissue - so if recombination can occur in one human, in how many others? The fact might seem to blow a whistle on the home run of Eve's game. In 1996 a study said: 'The current view (that there is no paternal 'leakage' of mt-DNA) is incorrect. In the majority of animals - including humans - the mid-piece mitochondria can be identified in the embryo even though their ultimate fate is unknown ... The missing mitochondria story seems to have survived - and proliferated - unchallenged ... because it supports the 'African Eve' model of recent radiation of *Homo sapiens* out of Africa'.[138]

Debate continues. Male mt-*DNA* does persist in the oocyte of some organisms and in a range of others different methods to destroy it are

---

[134] Maynard Smith J. *et al.*: Science 286 (24-12-99 ps. 2524-5); *ibidem* Strauss   p. 2436; and Zouros *et al.*: Nature 359 (1-10-92 ps. 412-14).

[135] Ross: Scientific American October 1991 p. 32

[136] Maynard Smith, J. New Scientist 14-6-2003 p. 50

[137] Schwartz, M. and Vissing, J. Paternal inheritance of mitochondrial *DNA*: New Engl. J. Med. (2002) 347, 576–580.

[138] Procs. of the National Academy of Science, USA 93 (24): 26-11-96 ps. 13859-63.

employed. It is known that both human sperm mid-piece and tail (with mitochondria) enter the oocyte but are then, it appears, destroyed - but *how* and how effectively? For mammals there exist two main theories, dilution and degradation. In the former sheer quantity of female mitochondria swamp any of effect of the male few. In the latter, called spermophagy, sperm parts other than the critical nuclear DNA are detected and marked by a protein, ubiquitin, for subsequent destruction. Ubiquitin's ubiquitous work in all parts of a body is attaching to obsolete proteins to signal to the cell that they are ready to be disassembled. *Why* male mt-*DNA* disassembly happens is another unanswered question. Is strict matrilinear inheritance therefore a theory that is going to survive?

Thirdly, it is assumed without evidence that chance mutations 'tick regularly'. In truth, however, how do you know how often human nuclear or mt-*DNA* mutates? Lacking any real clue you assume a calibration point. You adopt the prior and primary assumption of evolution. You then reckon, by the evolutionary interpretation of fossils, the point to be 5 or 6 mya. This sort of argument, first assuming what you wish to prove, is deeply circular. And there's more. You now assume that contemporary chimp *DNA* approximates to the genetic constitution of original Eve. Assumptions pile upon assumptions like a house of cards. If any one is incorrect then your construction falls. Anyway, 5 million years divided by the number of mutations from the assumed calibration point affords you a mutation rate, a molecular clock. Now, hopefully, you can estimate Eve's age, separation from mt-*DNA* group L0 and the whole phylogeographic tree. But have your assumptions, bound in gross circularity with what you want to prove - that man evolved from such a common ancestor - really worked?

The mutation rate may actually be up to twenty times faster than change from such a theoretical point requires. In other words, the time-scale for divergence may be dramatically reduced.[139] *Eve might have existed only ~10 kya.* This could, as the previous page implied, easily kill off any chance of chimp-human evolution. *And again, since we don't know what degree of difference might make a human not a human, we cannot distinguish between Neanderthal as different from, or the same as, our own species by this method.* For example, some Australian *H. erectus* bones have yielded mt-*DNA* like ours; so has anatomically similar if not identical Cro-Magnon man. On the other hand, an anatomically similar 40 kyr Lake Mungo specimen and some Neanderthal bones have mt-*DNA* unlike ours. Aren't they all actually human?

At this point we can interject that stories of humanity vary considerably but a current, rough agreement is that humans (*Homo*) with our brainsize and anatomy were present by ~230 kya. Yet no signs of

---

[139]   Gibbons: Science 279 (2-1-98 p. 28).

even crude 'modernity' appear until ~60 kya; and none of civilisation until ~7 kya. History knows how far our race has progressed (but not evolved) in those few years. 200ky is a long time for people as intelligent and cognitively able as ourselves *not* to learn to build houses, towns, roads and boats; and not learn to read, write and create sophisticated art, metalwork, engineering and so on. Do we explain this, as we used to in the case of Neanderthals, by stupidity? Or perhaps, 'impossibly' and 'unacceptably', the dating methods are wrong (*see* footnote 140)?

Anyway, assume that mitochondrial mutations tick a constant rate and so indicate, by means of this 'molecular clock', a chronology of changes. In 1987 a study of 136 women from many different racial backgrounds seemed to indicate a single individual or individual group with aboriginal (L0) mt-*DNA* dwelling in Africa about 200 kya. Further analysis, this time of European stock, seemed to identify an ancestry leading from five 'daughters of our Eve'; and the original human group may even have been as few as a Biblical-approximating two.[140] How can this be? Weren't there lots of Eves whose lines, except our own, have through conflict, illness or catastrophe, all gone extinct? At any rate, by some speculative but unexplained and unsubstantiated 'bottleneck', five or less women are conceived to have borne the indigenous children of Europe and the whole of imperial white man's worldwide commonwealth. A few more, it was estimated, bore all Asian stock as well. **From perhaps a single matriarch there issued everyone.** By all accounts, then, what a close and recent family relationship all humans actually have.

What serendipity created a 'genetic bottleneck' by which we all have, luckily, one Mum? Did the population crash by pestilence or waterless condition on the exodus from Africa?

It turns out that the computer program into which the mt-*DNA* data was loaded was 'order sensitive'. Thousands of subsequent runs in randomly different orders have not preferred an African origin. They have equally randomly selected its location so that nowadays the frank admission is that 'Eden' could be almost anywhere. It just depends on your authority's belief. Moreover, perhaps 'mixing' between tribes of different genetic constitution blurred the clarity of a single line to an extent that makes analysis of an original Eve impossible. It is not, of course, necessary that maternal and paternal mt-*DNA* might somehow actually recombine;

---

[140] The maths concur. If 'Adam' and 'Eve' were an original couple then, at an average of ~1.135 reproductive female offspring per couple thenceforth they could, over 10,000 years (400 generations of, generously, 25 years each) have theoretically produced the world's current population of 8 billion offspring. Such a low rate could be assumed to include reducing factors such as natural catastrophe, infertility, pre-reproductive death (as child or spinster) or smaller family size.

multiple types of quite dissimilar mt-*DNA*, from different parts of the known mt-*DNA* phylogeny, are reported in single individuals. However, in view of re-analyses and corrigenda of forensic mt-*DNA* data it is also asserted that 'the phenomenon of mixed or mosaic mt-*DNA* can solely be ascribed to contamination and sample mix-up'.[141] Mixing, no mixing, what kind of mix-up? Could a series of assumptions, born of desire to map a scientific Eve's career, become a mix-up that obscures her origin?

In short, the apple of Eve's metronomic tick is, based on circularity of argument, arbitrary and quite likely incorrect. Indeed, if the assumed regularity of a molecular clock (which is supposed to click in time with constant and yet, contrarily, random mutations) runs faster than imagined, it would make Eve's earth-time proportionately much nearer ours. How do you calibrate a clock like that without assuming what you want to find - evolutionary time-scales lifted from assumptions over evolutionary divergences between organisms? If these connections aren't correctly made then a densely-built conceptual network fails.

Do molecules and fossils fight? If a single Eve or troupe of Eves, offspring of 'Our Earliest Eve', emigrated out of Africa from (estimates keep varying) between 65 and 150 kya to populate a vacant space then what are Neanderthal remains dated 800 kyr (Gran Dolina) and 450 kyr (Petralona) doing in Europe? Also Dmanisi, Denisovan and *erecti* fossils further east aged up to 1.7 mya? Their presence begs the question where they first came from. Did Eve's tribe grow to wipe these isolated, antique populations out or, war and genocide apart, did they link up? Indeed, the Qafzeh cave remains (~100 kyr) indicate that humans of Neanderthal and modern type co-habited for up to 60 kyr.

Volumes of data are useless without correct and competent evaluation of their informative value; and this value depends on the scheme into which you load them, in the light of which you seek to find sense. Thus molecular comparisons applied to confirm the scheme of Darwinian evolution may, despite their raft of delicate assumptions, seem convincing until their first principle, the circularity of evolutionary presumption, is addressed. To this it needs be added note of difficulties that the analysis and interpretation of mt-*DNA* results involves.

Firstly, *DNA* is a very complex molecule and, as such, strongly subject to entropy. Causes of degradation include heat, pressure, time, dampness, oxidation, background radiation, microbe attack and so on. Its half-life is thought to be about 520 years so that, with optimum preservative conditions (say, dry bone at a steady -5 °C) its information would become wholly garbled in a million or so years. Under normal, natural conditions it would disintegrate in the wink of a geological skull - at maximum 10 kyr.

---

[141]   Journal of Medical Genetics 27-5-2005 (epub version).

However, a technique called polymerase chain reaction (PCR) is able to amplify tiny fragments into quantities that make them accessible, identifiable and analysable. This method is now universally employed .in genetic studies. As regards human evolution, in 1997 Svante Pääbo recovered and, despite the degradation which 40 kyr should have exacted, analysed mt-*DNA* from the 1856 type-specimen Neanderthal fossil found in Feldhofer Cave. Differences between this and contemporary samples have been interpreted to show Neanderthal divergence from our line 600 kya; and that interbreeding did not, against the fossil evidence, occur. However, using sequences retrieved from Sima de los Huesos in Spain and a toe-bone found in the Denisova Cave in Siberia, by 2013 the whole Neanderthal genome had been sequenced. The lads and lasses were calculated to have been ~ 99.7% human. This means that, since the modern human spread is 0.5%, they could mix with us today. *In this respect it is worth emphasising that nobody knows how different a genome must be in order that its phenotype is judged to constitute a new species.*

However, PCR results are susceptible to 'contamination', that is, the introduction of modern *DNA* into a sample by chemists. The slightest sample from the chemist will be magnified. The problem is: if as a precaution you then eliminate all 'modern' sequences from your PCR results (on the basis that they might represent such 'contamination') obviously you only retain what is different. Indeed, Neanderthal *DNA* too close to human would, under such a regime, be excluded from consideration; true Neanderthal material could, because of its *sapiens* proximity, end up being rejected. You would end up only amplifying 'different' material; and you would thereby, due to a methodological tautology, simply give a false impression of genetic distance and so 'prove' the *neanderthalensis | sapiens* distinction you might have theoretically suspected and set out to prove! You might, conclusively but also erroneously, proclaim Neanderthal separate from us.

Thus some experts despair of distinguishing between such close *Homo* genomes due to the great susceptibility of their method to contamination.

A similar result occurs if you use PCR primers that will only amplify Neanderthal mt-*DNA*. You may thus find a particular stretch of genome absent but this proves no more than methodological bias, an incomplete sample or genetic variation in your individual case due to drift, 'bottleneck' or other factor. *The real issue is the arbitrary exclusion of modern material, presumed due to contamination, which might actually be from the Neanderthal.* Also, ancient mt-*DNA* may have undergone significantly more mutation than nuclear. Neanderthal *nuclear DNA* might well fall within the orbit of modern human while, on a faster timescale, an mt-*DNA* sample seem too divergent.

To recap, the picture is disputed and unclear but it would seem that African Eve theorists believe modern humans started in the same place but it was not Eden. Instead, you might allow gradations of morphology in Africa; some evolutionary continuity should build towards humanity. Eve's hypothetical troupe then sallied out of the dark continent to claim (perhaps several times) a wide, un-promised, intercontinental land. Here you would expect discontinuity, a sharp distinction marking out 'superior' conqueror from 'inferior', vanquished and sub-human stock. Did Eve's people slaughter all 'inferior' specimens they met? There was, adherents claim, no ethnic interbreeding with Neanderthals or *H. erecti* either west or east; they were just replaced. This is not, however, what the fossil record demonstrates; nor, today, do people round the world resemble African replacements.

Thus the story of humanity ex-Africa (or, you might add, ex-Eden too) involved both incest and then fratricidal war! This marks man's vector of ($\downarrow$) descent into the world and not ($\uparrow$) ascent detaching from it.

How much, therefore, agrees with theoretical predictions? At most basic, is 'Our Lucky Lady' real? And did her daughter's Exodus, pregnant with the greatest consequence of any human emigration, actually take place? Or is Eve a fiction born in California a few years ago?

'Mitochondrial Eve' adherents have developed plausible but impossible-to-prove assumptions involving molecular clocks, constant rate mutations, date-calibration points and the reconstructed genetic constitution of 'Our Lady' (lived ~200 kya in South Africa) and an 'Emigrant First Lady' (lived in Ethiopia say 100 kya). They also assume the integrity of strictly matriarchal mt-*DNA* time-lines, no mixing with non-African natives, uncontaminated PCR analysis, the employment of circular arguments and, above all, intransigent taboo against the interpretation of any data except according to 'the fact' of evolution. Assumptions pile like a house of cards.

This said, did the earliest humans hail from Africa? Jebel Ihroud, Idaltu, Rhodesian and Saldanha Men are African. But other fossils, Eve's elders too, are spread right across Eurasia from Spain, England, Germany and Armenia to India, Siberia, China, Indonesia and Australasia. Indeed, high levels of Denisovan genome are, as already noted, found today in Papua New Guinea and Australian aboriginals. Experts currently agree that the accumulated weight of oriental bones has forced the narrative that bridges mankind out of Africa into collapse.

To and Khoisan fro. No doubt, slung from rare slivers of bone elaborate hypotheses are swung this way or that but very often lacking sure evidence how different species, races or varieties of humans *must*, as theory instructs, have evolved from any other or progressed from apes. *If, on the other hand, you count erectus, 'archaic sapiens',*

Neanderthal and sapiens inside the human archetype you don't need any special exodus from Africa. **Isolation, small groups, allopatric speciation, genetic drift and adaptation from programmed potential ring continual changes so that, as today, variations over time occur. This, micro-evolution, is not macro-evolution. <u>Variation is not transformation</u>.**

Experts, locked into theory that inherently demands they must, split up rather than unify the human type. They speciate and in profusion classify but it seems clear that Eve's supposed colonialists did not wreak genocide on Dmanisi, Denisovan or other population pockets of Neanderthal. Sima de los Huesos ossuary in Spain and caves in Israel have also testified to interracial mix. Sub-species of the human family survived in harmony for, at least in the case of Amud cave, 50,000 years. In this natural vault a Neanderthal skull (Amud 1) was found buried just below the surface of a layer (called B1) which has been carbon dated to 5.7 kyr so the skull can't be older and may be younger than that. At this age it is academic whether I am called '*neanderthal*' and you '*sapiens*'. Say negroid, Aryan, black, yellow, pink or white. We lived together until practically historic time! At which observation, faced with theoretical anomaly, an expert soon steps forward and, on behalf of Darwin's theory, pronounces that contamination made the sample seem too young! An exhibit that opposes theory has been 'falsified'.

23.2   Idaltu Man ('photofit'): see also *fig.* 19.6 for actual skull

What happened happened. Perhaps a trail from Africa was blazed. Space-time convenient fragments (three broken crania, other pieces of skull and some teeth, ~160 kyr) found on the road through Ethiopia and called *Homo sapiens idaltu aka* Herto Man, can be arraigned to demonstrate the case. Yet, if you are not constrained by thought of progress in an evolutionary sense, you understand with ease that humans always have explored. They travel to and fro across the world to trade, exchange ideas and improve their living standard. It is called 'upward mobility'. And mobile voyages don't have to take a million, a hundred or even a single year. Indeed, fossils and tools of 'archaic' men, as modern as Idaltu and dated ~300 kya, have been found at Jebel Ihroud in Morocco. Africa *may* have been the centre humans sometime radiated from but there are other possibilities. Victorians thought that Europe was the place. The Middle East and Orient are candidates as well.

What happened happened but, of course, speculation and abductive reason[142] *are the only ways to peer into the past.* **The present may or may not tell you what occurred but no-one can, experimentally, roll back prehistory.** As regards molecular interpretations this also applies. Phylogeny, phylogeography and large-scale evolutionary guesswork demonstrate computer simulations are all fraught with formulaic assumptions and perspectives weighed with differing hypotheses. The 1987 study of 136 women's mt-*DNA* wags a salient warning. In this case, we saw, the program was constructed on false premises. As well as Africa Eve could have come from practically anywhere. This discovery provoked an editor at Nature Magazine, Henry Gee, to slate its 'Out of Africa' proclivity as 'garbage'. Garbage in, garbage out. All depends upon your frame of reference and, thence derived, assumptions.

In the end, in a letter to Science,[143] one of its founder theorists, Mark Stoneking, squarely in the scientific way admits fresh facts may have overwhelmed the 'Mitochondrial Eve' hypothesis. Furthermore, in late August 2017 news broke that 'hominin' footprints had been found at Trachilos in western Crete and securely dated 5.7 mya.[144] Feet are a defining characteristic of humanity and these fossils are precisely human: but their Miocene age is about three times earlier than previous speculation regarding our beginnings. **Therefore they shred not only the out-of-Africa fabrication but the whole, fragile palaeontological tapestry painstakingly woven over the past century and more.**

Workers such as Bräuer won't, for now, agree with Stoneking and thus split species and play down the other models that we'll now review. More seriously, the fifty footsteps leading Darwin's theory into wilderness must be accommodated. So hold on. In a creation narrative so flexible and open-ended will the next step be the evidence somehow explained away or, within materialism's evolutionary framework, new explanations altogether tried? Because, if evolution is a fact, there'll be a way that's plausible to herd the facts into its story-line. Just-so.

---

[142]   see Glossary: logic.
[143]   Science 255; Feb. 92 ps. 737-739.
[144]   Uppsala University Press Release 31-8-2017 and New Scientist 4-9-2017; see also *Graecopithecus* (Chapter 20) and Laetoli footprints (Chapter 19).

24. *Adam and Evolution*

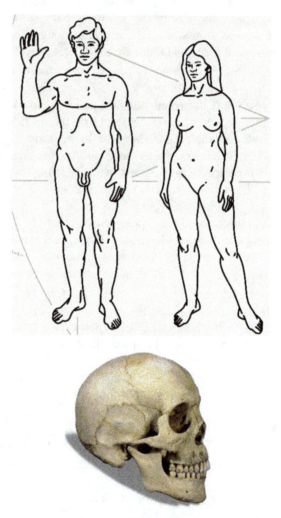

24.1 This is you.

Rival Suitors

This is you. And theories of your origin make ebb and flow. Band-wagons trend and trend again. Eve from Africa is one. Has this First Lady rival suitors? And in principle, could mankind have been conceived in any womb but that of 'Our Fashionable Madonna, Lady Luck'?

Marcellin Boule and Sir Arthur Keith proposed that *H. sapiens* evolved from Australopithecines in parallel with heavy-bowed, robust *erectus* and Neanderthal. This idea is long *passé*.

In the 1930's Franz Weidenreich proposed a **Multi-regional or Polycentric Theory** of human evolution wherein *H. erectus* evolved into *H. sapiens* independently (at different places) across its whole Eurasian range. However, a mobile network of lineages joined by 'gene flow' ensured that useful characteristics spread from one nexus to the others and also that, with no pockets of *erectus* left untouched, uniquely *sapiens* emerged from the process. And the races of mankind were simply the result of regional variation.

This reasonable, **polycentric idea** was elaborated by Harvard academic Carleton Coon in two books - **The Origin of Races** (1962) and Living Races of Men (1965). He proposed an evolutionary model involving five main human races - Caucasoid/ Aryan, Mongoloid, Capoid (bushmen), Congoid (negroes and African pygmies) and Australoid (Indian and Australian aborigines, New Guineans and south Asian negroid pygmies). Each was, he thought, derived separately from widely dispersed populations of *H. erectus*. He placed less emphases on a network of 'gene flow' than Weidenreich and believed each group may have started to evolve at different times. He estimated, for example, that Mongoloids had been evolving since Java and Peking Men (say, 700 kya), Europeans since Neanderthals (say 800 kya, Gran Dolina, 500 kya, Heidelberg Man or 450 kya Petralona Cave), Congoids from (as in 1962 he dated Rhodesian Man) 40 kya and Australoids from an even more recent date.

However, in the 1960's times they were a-changin'. The biological theory of evolution is, as we've seen in Chapter 7, intrinsically a 'racist' (strictly, 'fluid species-ist') theory. The fittest dominate, inferiors die out. From this perspective the differential start-dates and, possibly, rates of evolution in Coon's model implied that some races were more evolved, that is, fitter than others. However, in the post-war era western culture with good social conscience if not evo-biological rationale excoriates the turpitudes of genocide and slavery. Aren't men born equal? This was, moreover, the decade of the U.S. civil rights movement, Black Panthers and so on. Not only was the model caught in the flak of an intellectual feud by Columbia University's Franz Boas and Ashley Montagu (whose approach was emphatically cultural rather than biological/ morphological); it was also usurped by segregationists who, as Darwin had been usurped by Nietzche and the Nazis, claimed that ex-African blacks were 'American juniors'. Wherefore, Coon's pact with fact now whiffed politically incorrect. An academic quest became embroiled in a socio-political firestorm.

Since when have sociological theorists and political activists cared too much for scientific truth? Nevertheless, many of the scientific community, showing their moral conscience, veered with the consensus. Herein flexes tension between science and morality. You cannot blame a theory for the sins of man; you *can*, however, deprecate the fact that Darwin's intrinsically justifies them.[145] A more conscionable, unifying and 'politically correct' theory was demanded. And Coon was ditched.

Rather than replacement of Neanderthals by 'Eve's People' Smith and Trinkhaus have blurred the image of strictly separate species by allowing, in their **Assimilation Theory**, that interbreeding was the norm. Diffusion and not domination is its half-house theme.

Now let's take a fourth shot at conceptual ape → man flow. Milford Wolpoff (USA), Wu Xin Zhi (China) and Alan Thorne (Australia) have, by extending the ideas of Weidenreich and Coon, promoted a **Multi-regional Continuity Model**[146] characterised by extensive migration flows and interbreeding/ hybridisation In this case *H. erectus*, some of whom emigrated from Africa ~2 mya, evolves to *H. sapiens* independently in Asia, Africa and Europe. Although there were no more singular, Eve-type excursions from Africa, sufficient intercontinental 'gene flow' prevented the production of three continent-specific species of human. You might dub this kind of evolution semi-parallel. Therefore, oxymoronically for a 'non-racist' crack at explanation, the MRC's search for truth is, like Coon's, still open to the charge of 'racism'. To escape this charge is, of course, why 'Mitochondrial Eve Theory' posits a single ancestral stock, evolution over the same period and thereby recent, shallow differences between all humans. It is also why proponents try to distance us from Neanderthals, already present in Eurasia when Eve's Exodus occurred.

Because no specific site for *erectus* > *sapiens* evolution is identifiable the only separation point left is time. However, is time alone a sufficient agent of progressive speciation? Is it even, remembering the ancients and modern two-named alligator of Chapter 21, sufficient to split a species into two? **MRC decides it is not and therefore excels in classifying *H. erectus*, *neanderthalensis*, archaic and modern *sapiens* as, in fact, just *Homo sapiens*.** Thus you find *Homo sapiens neanderthalis* in the books; and *H. s. erectus*, *H. s. heidelbergensis* and (don't worry) *H. sapiens sapiens*. In this case there's only *H. sap.* interest for at least a million years.

Why not? Dobzhansky and Mayr's Biological Species Concept (BSC) defines a species as consisting of populations of organisms that can reproduce with one another and that are reproductively isolated from other such populations. It also needs be recognised that species concepts are humanly produced categories which may or may not always work when compared with the reality of nature. Thus, is *H. erectus* so

---

[145] *SAS* Chapter 26.

[146] Modern *Homo sapiens* Origins: A General Theory of Hominid Evolution Involving the Fossil Evidence from East Asia (1984).

morphologically different from *H. sapiens* as to warrant a label of 'incompatible species'? In which case you might categorically forbid his presence as a variation possible by genetic recombination, interbreeding, isolation and adaptive super-coding.[147] Or is he found in a time-frame that permits legitimate transitional forms, by mutation, natural selection and evolutionary genocide of less evolved competitors, to grade into modern man? By now the *erectoid* designation, a slightly smaller version of Neanderthal, is composed of over 280 fellows worldwide. Their dates range from ~6000 years (Mossgiel and Kow Swamp crania in Australia), *erectoid Homo soloensis* (~35-53 kyr) through Peking Man (300 kyr), Java Man (skull, 700 kyr, with human femur and nearby Sondé tooth), various Olduvai hominids, Tanzania (~ 0.7 for OH 22 - 1.9 myr for OH 60), Swartkrans fragments (1.8 myr) through to cranial (KNM-ER 2598) and hipbone (KNM-ER 3228) fragments, Kenya (1.95 myr).

What, moreover, of Dark Eve's place in Sima de los Huesos, Dmanisi or Israeli caves? What about the borderline morphologies suggesting human/ Neanderthal interaction across eastern Europe, Neanderthals dated 5.7 kyr at Amud Cave (surely some mistake, contamination perhaps) and, since they help exemplify this book's debate, Australian fossils?

From Kow Swamp (KS), surrounding area and other parts of Australia the fossils are generally agreed (although different methods such as carbon 14 and OSL luminescence give different dates) to be between 6 and 35 kyr. This immediately raises two problems, one relatively trivial and the other serious.

1. The normal paucity of material from which reconstructions, casts and photographs have been taken. Most, having been reburied at the wish of local aborigines, is now unavailable for further study. Such paucity leads to clouds of speculation. For example, the only relatively complete Kow Swamp skulls (KS1 and KS5) are arguably *erectoid*.

2. All parties agree that representatives of *sapiens* or 'archaic *sapiens*' comprise part of the assemblage.[148] They also agree on the use of tool, artefacts, fire and burial by the specimens. However, the proportion and dates of fossils diagnosed *erectoid* raise an unacceptable challenge to the notion of Darwinian succession; and

---

[147]   *SAS* Chapter 23: Super-codes and Adaptive Potential.

[148]   Some *H. erectus* remains from Kow Swamp, Australia have been diagnosed human (Proceedings of the National Academy of Science, USA 98 (2): 537-542); also, similarly, three Neanderthal and one anatomically modern Australian (Mungo Man 3, 40 kya) have mt-*DNA* different from contemporary humans. In this case, perhaps *H. erectus* bred with us and, across the other side of the world, Neanderthal as well - because they were us!

the salient anomaly is sharpened by a dating of 'robust', *erectoid* samples (e.g. KS1 and KS5 at ~5-9 kyr)) as younger than the 'gracile', *sapiens*-like ones e.g. Mungo Man (LM3) and Woman (LM1) at a disputed ~35 kyr. Where is the time for evolution of *erectus* into *sapiens*? Worse still, the anomaly implies that *sapiens* and *erectus* not only co-existed but cohabited.

Thus anomaly afflicts the 'Out-of-Africa' model, wherein *sapiens* exits Africa about 60,000 years ago and does not co-mingle with Neanderthals or *erectoids* derived from previous emigrations. It also severely compromises the Darwinian principle of succession. As such it provokes two or three typical reactions. Firstly, Academia and the Press (even, uncharacteristically, in the form of research-funding, evolutionary-hyping National Geographic) prowl the edges but, by disregard or omission, stay clear. Secondly, the less circumspect and self-styled 'anti-creationist' contingent launch into their usual 'hakka', a two-step war-dance which pronounces everyone who disagrees with Darwin 'creationist' and, next step, binnable. Thirdly, hot dispute, often involving bone-headed inflexibility, occurs.

Therefore, for example, welcome in the evolutionary corner palaeoanthropologist Dr. Peter Brown and his second, Jim Foley of 'anti-creationist' website fame. Their objective will be to deliver KO for ideas of very late (and thereby anomalous) *erectus* presence. At the whistle Foley requires Brown to box clever by comparing KS fragments with 16 characteristics generally used to diagnose *H. erectus*. In general his tactic is to admit that, while some specimens are 'gracile' moderns, the others exhibit more 'robust', 'archaic' features (you may remember Broken Hill Man, *H. heidelbergensis* and others from Chapter 19). Thus, while its boundaries may be blurred, the KS collection is human without *erectoid* 'contamination'.

On the other hand, welcome counterpoint from another popular 'origins' blogger, Jim Vanhollebeke (whose cornerman may or may not be Martin Lubenow). Vanhollebeke parries each of Brown's 16 punches. One equally notes the evidence of *Homo erectus soloensis* and, specifically, Sangiran skull 17 from Java. Franz Weidenreich commented that *soloensis* characteristics were found in quite modern aboriginal material and Garniss Curtis dated Solo Man at ~35 kyr. Thus a following right hook might muscle in with evidence of *erectoid/ Neanderthal* and possibly *erectoid/ sapiens* co-existence (as with aforementioned Amud Cave, Israel and Sima de los Huesos, Spain). Co-existence would normally mean, *inter homines*, cohabitation - hence, for example, Neanderthal appearance in our genes. No KO for Darwinism's corner here.

The upshot of the match typifies the thrust of this book.

There is a *bottom-up*, materialistic view that evolution is our truth and facts must be interpreted within the theory. *Indeed, without a theory of evolution there is no materialism nor, in that sense, 'scientific explanation' of life's origins.* Since life is an integral part of things, the mind-set and philosophy downright cleave to and depend on it.

There is, conversely and despite materialism's '*hakka*', a *top-down* informative, archetypal perspective. This submits that, from a holistic angle, the human type (*Homo*) includes Neanderthal, *H. erectus* and *H. sapiens*. *This means that these classifications mean as little as racial classification today.* **And that the three are variants in time (as there is much obvious variance in terrestrial space) of a single type, the same basic, human archetype.**[149] Fossil relics simply manifest adaptations and differences due to adaptive genetic potential, random drift, allopatric speciation and the general isolation that small groups of humans are bound to experience across the vast face of earth. This conclusion, because it renders ape > man speculation (deriving perforce from a Darwinist angle) vapid, is of great impact and unpopularity within the academic, scientific human evolution 'industry'. **No-one welcomes the collapse of their investment in Authority but, if the basic premise of an undertaking is flawed how can its conclusion be correct? In this case philosophical position, not evidence, is the only reason for ape | man rejection.**

In short, the MRC model has the merit of 'sinking' interbreeding *H. erectus*, *H. neanderthalensis* and *H. sapiens de facto* into the same species; nor, due to interbreeding, would three separate populations of *H. sapiens* evolve at different rates and times; and the single species would show a measure (such as we know today) of geographical, genetic and morphological variation. The theory does, however, still require the usual, extended periods of evolutionary time. Both Eve and MRC models date *H. erectus*, despite contrary evidence, from between 0.4 and 1.5 mya; but at least 140 specimens are younger than 0.4 and over 30 dated older than 1.5 myr. The latter are, apparently, ignored. The slightly taller Neanderthals are, in turn, dated from about 400 to 40 kyr. During much, perhaps all, of such a span either modern or 'archaic' *H. sapiens* was, it seems demonstrably, alive (*cf.* Laetoli 3.75 myr and Kanapoi 3.5 myr).

Neo-Darwinism is a theory riven with gaps and guesses, millions of them with, as we've seen, a fair share of them when it comes to humankind. None of the three or four brands of hominin evolutionary theory solves this problem of extended parallel coexistence that the actual fossils puzzlingly, perhaps even irritatingly, display. Maybe

---

[149]   see Chapter 9: Limited Plasticity.

(though fans of 'Mitochondrial Eve' and MRC will bitterly object) there's a better way to blot out racist implications that evolutionary theory holds for the ascent of man.

For example, what triggers an individual's *DNA* that code for brow-ridges (or the lack of them)? What adaptive potential (Glossary) might trip certain types of characteristic to better thrive in long-term, extreme conditions (such as an Ice Age or, more locally, a scorching desert)? And what switches code for different skull sizes in a person (*fig.* 21.5) or an isolated population? We don't yet know but it is, despite the MRC model, certainly unfashionable to lump together similarities let alone chronologies between Neanderthal, African (Idaltu or, say, KNM-ER 1813), Georgia (Dmanisi, say, D2077), Indonesian, Denisovan (Russian), Spanish (Sima de los Huesos) and (Chapters 17-19) Chinese skulls, each sample in its paucity representing populations. Such 'anti-splittism' is not good for abundant species labelling or devising new phylogenetic lineages. An international, pre-technological community, if nomadic occasionally interbreeding in a sparsely populated world, is not a suggestion that the expert mind-set welcomes. It severely sprains lifetimes of 'mainstream' custom dedicated to arranging facts by theory. Yet if you, also a hominin, were placed in such prehistoric context (as some stone-age dwellers in the world still live), would you be judged sub-human as Darwin initially judged the Fuegian 'primates'; or caged next to Ota Benga in Bronx Zoo? How dumb and brutish would you paint friends' expressions, how foolish would you find them?

However, with recognition of *Sapiens/ Neanderthal* interbreeding (Chapter 14) the pictures are modernising.

24.2   Not quite you. *H. longi* (skull and artist's illustration)

Recently another human sub-species, one with a *Sapiens/ Neanderthal* cranial capacity (~1425 cm³) and thought perhaps to be related to Denisovans (*H. sapiens ssp. denisova*) popped up 'in mysterious circumstances' (New Scientist 25-6-2021) with all the usual media fanfare and nicknamed Dragon Man, from a well in Harbin, China. This chap, found at the Songhua river, had laid hidden since 1933 because the peasant finder did not dare tell even his own family until recently when he was on his deathbed. The skull is complete except for teeth and mandible (a pity since no comparision can be made with similarly-aged Denisovan (?) mandible with teeth from Xiahe, Tibet. No *DNA* analysis has been performed yet, but beetle-browed Dragon Man (*H. longi*) may, by U-Th radiometric dating, have lived 0.14 ma (140,000) years ago. Despite several other similar, approximately contemporaneous Chinese skulls, for example, 'archaic' Dali, Dragon Man is not thought to have interbred and his line is extinct line is considered extinct.

First Principles

Back, therefore, to first principles. A materialist does not believe an immaterial element exists. Since brain is mind, he counts reason, logic, mathematical calculation and symbolic conceptualization as material things. Non-living matter experiences sensation since his world is not symbolically conveyed to him. Non-conscious protons and electrons make the thoughts he thinks. Feeling is, since matter makes only material things, an unexplained excrescence from his nerves. Man's mind is grey jelly - but can the jelly explain how? In short, matter maketh man whose jelly head is all there is to mind. All palaeoanthropology (as well as science and biology) is based on this assumption. All speculations roam accordingly.

This roaming is because physical science wants to understand and seeks, behind the apparent chaos of this cosmos, order. If, however, chance is order's opposite then to propose that order's source is chance appears perverse. How can you ever pin down randomness or build integrated texts from it?[150] Thus any theory proposing chance has caused the accurately coded order of our bio-world is, at best, unrepeatable, undemonstrable and, practically, a narrative composed of 'what ifs' and improbabilities. Its staple is a 'must have' couched in forbidding, combinatorial space. The theory of evolution is just such a theory and proposes, as materialism must, that the source of bio-order sums to chance mutations coupled with environmental stress. This is not, therefore, except in its rejection of an immaterial element, a scientific theory. *There are Theories of Intelligence and No Intelligence.* **A theory of the latter, invoking as its driver chance, is no more scientific than the former viz. a theory of immaterial, informative intelligence.**

If, however, you accept an immaterial element of information you reject materialism's evolutionary premise; you intrinsically accept *a top-down* point of view. In this case[151] Top Teleology involves, at root, Pure

---

[150] *SAS* Chapters 7 and 8.
[151] *SAS* Chapters 1-3, 5 and 6.

Immateriality. Not pure, non-conscious energy (like light) but a Perfect Balance of Potential Information. Essentially there is Nothing but this One. Motion of the Quintessential One breaks balance; it creates existential (↑) ups and (↓) downs, a field of relativity. Concentrated Information, moving, becomes mind. Ideas are purposely conceived. Mind moves and in principle programs are drawn up - *but no physical creation yet exists*.

In a computer files are stored. The natural store is memory. Mind plans projects; its actions (and impressions) are stored naturally in the subconscious state of mind. Subconscious mind is nothing if not files of memories; who lives in instinct or in habit lives, effectively, in memory. And universal files in universal mind are labelled archetypes.[152] From there, as any mind projects its plans, cosmos is projected; physical mindlessness and, within the order of its rules, purposeless contingency now form a wholly energetic universe. And on that cosmic stage life-forms partake, in varying degrees, of conscious and subconscious mind. Humans, you and I, are in that group.

Symbols (such as words, numbers or pictures) may be employed to passively reflect or actively innovate physical objects or events. Nor is their immaterial code the same thing as material reality. A computer program and its files are not the same as pictures on your screen; nor is a specifically-codified genome the shape of the organism that it represents. But a code (e.g. an intelligently automated car production code or bio-productive *DNA*) that can realise its story, that is, materially produce the meaning it encodes is quite astounding. Active symbol precedes, passive follows physical events. Immaterial, informative potential precedes its target outcome. On a grade of intelligence it clearly scores 1 where (even if you allow a ratchet holding 'accidentally correct' hits and called natural selection) chance at 0 is right out of question.[153]

Thus to the crunch.

Bottom-up, ape > man. It's not, of course, an evolutionary conclusion that Adam and Eve started life together. It depends, in gradually shifting sands of species-change, on where you draw the 'human' boundary line. We now think that males and females from the world's '*cluster*' of *hominins* may have mated with so-called different species for 50 kyr. We have evidence in our own blood. So how can the mt-*DNA* and Y-chromosomal estimations date the start of man? All are guesses where and when the shape we recognize as human first occurred.

Top-down, it's ape | man. *Why, naturally, should this rule out a possibility of extinct, 'allochronological' variations on either chimp or*

---

[152] *SAS* Chapters 16 and 19.
[153] *SAS* Chapters 20-23.

*human theme*s? However, nobody knows how sexed Adam and Eve, as we name the human projection,[154] were *initially* co-cast from archetype to physicality; nor can anyone prove where and when.

To repeat: mind is prior to matter. This, the revolution of materialism, is that faith's anathema. *SAS* has, *top-down*, challenged that anathema and logically demonstrated that, if just an immaterial element of information is allowed, it follows mind is precedent and physicality resides therein. Metaphysical precedes material and information runs the show. The rationale is, for the thorough reader who has followed through, an impressive one.

Most of us assume the universe began with a projection of some kind from nothing - certainly a nothingness beyond the laws of physics. Could not complex codes for living forms have been projected from material nothingness as well - an original expression of our physicality that stems from a conceptual archetype in universal mind? In other words, could not the seed of man's descent have developed, in a way reflected in the working of his own comparatively dim inventive

---

[154]   *SAS* Chapters 17 and 24.

genius, from a metaphysical idea into its physical elaboration?[155] The race of men believed it. Still most feel it right but today materialism, with a different story of creation, certainly does not. After all, it sounds as much a miracle as big-bang[156] or a proto-cell by chance![157] At least, the *top-down* explanation is a rational one!

Behold, therefore, two pillars of faith. *Holist or materialist, who in the end analysis is right?* Is it Adam and Eve or evolution of the apes?

Libraries of volumes expound the chance-ridden evolutionary hypothesis. Some faithful readers tend to find alternatives unreadable. Nor, like communistic politic, does their religious arrogance brook competition. 'Do you believe in evolution? If not, aren't you 'creationist' (for which read an 'unscientific fool')? This shallow two-step, dancing for its own mind's sake, thence feels empowered to reject a sensible consideration other than its own. *Yet (who can believe it?) hallowed institutions, all scientific magazines and most media establishments indulge unwisely in this jig.*

Holistic thesis, equally, is crystal clear. No son of ape rose into Tarzan. Man is a distinct type. He did not diverge from an evolutionary trunk: his tree grew straight up from its own roots. Apes and humans existed in the past as they do in the present, as separate types with large but limited variability. Interpretations lifted from molecular research are, depending on perspective, open to debate. And Natural Dialectic, accepting a substantial immaterial element, might treat the skulls and bones this volume has discussed as once children of all men's Creator (while an atheist would not).[158] Material of great contention, it is claimed the fossil evidence substantiates holism's case i.e. that fossils proposed as 'a vital series of gradated missing links' to illustrate the evolution of mankind are in reality (when they are not frauds) either men or apes. They do not sit easily between.

In the last analysis, humans, perhaps created in the watered parks of India, the Middle East or even Southern Africa,[159] *must* have diversified

---

[155]   *SAS* Chaps. 5, 6 and 19; *PGND* Chap. 12; *SPFP* Chap. 2 and *RSP* Chap. 2.

[156]   *SAS* Chapters 7-12.

[157]   *SAS* Chapters 20-21.

[158]   *SAS passim* but especially Chapter 1.

[159]   In Hebrew, Turkish, Hindi and several other languages, **Adam** means 'man'. **Eve** may be from 'Hawa', a Hebrew proper name. Its meaning is less clear but is possibly connected with the word for 'life'. Thus Eve, a womb-man, gives birth; she is the 'life-giver'. 'Eden' in Sumerian means plain. The septuagint translates Eden into the Greek word *paradeisos*. This word is of Persian origin and means a watered pleasure-ground or hunting-park of the

and migrated out, as wandering tribes, to all the habitable corners of earth. *Did the early races or species of man look and behave like stone-age natives do today - whether bushmen, Aborigines or Amazonian Indians?* **How well would you be coping after a few months if you were 'loosed', with all remembrance of technology and civilization erased from your mind, onto virgin continent?** It would be a fresher start than you or I can imagine - an Adam or an Eve. If you and a mate survived you might generate peoples who, as they radiated further from the original centre, developed their individual technologies and cultures. After many an epoch some of these fresh locations and their civilizations might well come to surpass that of the society from which, millenia previously, they had emigrated.

---

kind that Xerxes and Cyrus devised. Such a natural paradise would represent a good start for human physicality, that is, for earthly life.

25. *Mythological Mirage*

Do you remember anti-parallel perspectives?[160]

From a *bottom-up*, ape > man perspective natural forces are the sole creators of material form. This is an atheistic world-view and, although it should not philosophically be, the scientific *modus cognoscendi* too. Thus chance-based evolution is, perforce, the answer to the origin of life.

Scientific methodology feels, of course, unhappy with the notion of an immaterial, innovative element that's out of its mathematical control. Such element is not, however, unfamiliar. How, for example, do you think of memory? It is stored in your subconscious mind. What exactly is it? Close your eyes and try to see.

From a *top-down* perspective archetypal memories are stored in universal mind.[161] Think of them as programs or as stencils lodged as files in what is called potential matter. Therefore, in the hierarchy of creation, the world's immediate creator is metaphysical. Similarly, forms of life derive from passive memories filed in a subconscious phase of mind.[162] Such information is materially expressed in a correlation we have labelled *DNA*. Life certainly depends on prior information. **From this perspective, therefore, despite evolution's protest movement you would query whether Darwin's theory were not misinformation as primary as mutation to bio-logical sense; and thus a vast and world-wide exercise in self-deception, a fog of illusion whose perpetual swirling veils the order of truth.**

Thus this book, with *SAS*, effectively drives the thin end of a philosophical wedge into materialism's scientific block. If men have waltzed away with evolutionary steps in concert they'll condemn its interruption to their own Victorian Composer's simple score. However, in no way can condemnation render, by its own volition, *top-down's* proposition, man | ape, incorrect or false.

What matters is the fact of codified *design* and the sensible inference of *deliberation* it is possible to draw from a study of the biological machines we call plants and animals. While it entirely accepts science and its methods, the holistic view has roots beyond the materialistic vision of science. 'A Mutant Ape?' and its mother volume, 'Science and the Soul' will thus, of course, be controversial. The many issues it raises cannot all be dealt with, let alone in depth, in a single

---

[160] Chapter 1; also *SAS* Chapter 0 and *fig.* 24.2.

[161] see Glossary; also *SAS* Chapters 15-17 and 19; also *figs.* 3.1-3.

[162] see *SAS* Chapters 3, 16, 17 and 19: Conceptual Biology; *A&E* Chapter 17.

sweep. But the direction of the argument is clear - there has been neither chemical evolution nor macro-evolution.

No doubt, this book has illustrated evolutionary imperative and skew. Howell's Parade sums up its central notion of our fathers in 'an up-and-coming' line. Top of today's pops shows, of course, mankind. By fluke the illustration has been made since any other fluke back down six hundred million years could have diverted 'progress'. Howell, you and I would not have been. All praise, then, to a trillion accidents that have, like angels, 'guarded progress' up till now. Oblivion's Lady Luck has given birth to humans.

Mutation, natural selection, adaptation[163] and so on. We know. Everyone agrees these processes occur. *Top-down*, **however, simply adds an immaterial element and this, informative potential, revolutionizes how you think about yourself and where you'll logically find your source.** If cosmic order, well reflected in your constitution, is derived from mind[164] then program, complex integrity and specific purpose snap to fore. Indeed, does not any genome (a most sophisticated coding mechanism whose digital sequences deal in millions of genetic switches and instructions) lay the ghost of Darwinism like a light?

Genomic logic is quaternary (more complex, using four nucleotides instead of binary on-off) and can express the same target using different algorithms. **Whence the informative instructions?** Why is a notion of the necessity of a systems analyst, engineer or programmer to generate a symbol (genome) representing end result (development of reproductive body) so toxic? *The answer is, of course, intelligence.* For materialism such abominable anathema, if true, would destroy institutions, associations, departments and careers that work researching evolutionary interpretations of the bio-facts; and worse, such 'creationism' would subvert the secular religions (humanism, atheism, scientism, communism and the like) that are constructed as if the ultimate species of reason was human! *No wonder that the obvious is religiously denied and mythological mirage preferred.*

In 1823 Founding Father, author of The Declaration of Independence and the USA's third president, Thomas Jefferson, wrote to the second, John Adams:

"I hold (without appeal to revelation) that when we take a view of the Universe, in its parts general or particular, it is impossible for the human mind not to perceive and feel a conviction of design, consummate skill, and indefinite power in every atom of its composition."

---

[163] see *SAS* Chapters 22: The Editor and 23: The Creator and Super-codes.

[164] *SAS* Chapters 2-11, 16, 17 and 19.

In other words, the argument from natural intelligence (or Natural Intelligence) is not based on appeal to religious revelation but empirical observation of, in all its intricate organization, the natural world. Such insight makes a farce of the intended insult 'creationist'; the issue of design is not a question of religion but logic; and its appreciation draws perception past the mythological mirage of chance as author of our days.

It is certainly possible, as shown in Chapter 23, to compute that current human population could have been achieved from a single Eve (not quite forgetting Adam) in 400 generations over, say, 10,000 years. Such mathematical observation apart, this book has indicated that time-span, either short or long, is not the issue. Informative instruction is. Even the remotest probability that, in the span of this universe, chance built huge and precise banks of information (genomes) which underwrite the operation of all cells on earth is *informatively IMPOSSIBLE*. **If so, the origin of man's descent is metaphysical. It resides in archetype.**[165]

Thereby, with the ever-immaterial element of purposive design, an explanation of life's forms (including men and apes) is eminently plausible using Jefferson's empirical common sense. In this case blind forces, random mechanisms and the unexplained instincts of self-reproducing chemical incorporations to outwit entropy and thus survive are supplemented. *Metaphysical (or immaterial) archetype informs physical bio-design.* **The latter's genomic program allows symbol to construct, amazingly, its own reality.** This reality includes, therefore, adaptive potential[166] with its super-coding that supplies foresighted flexibility. Conditions that can trigger local variation in the form of limited plasticity include isolated population (by whatever cause), genetic drift and allopatric speciation. Adaptations are intrinsic in genomic type. If this simple thesis renders Darwinism dead then the extinction means that ape > man goes extinct as well. And Natural Dialectic's apposite suggestion, ape | man, is meritoriously sustained.

To summarise, one thing is certain. Through all the muddling and dispute every fossil is habitually interpreted *only* through the 'correct' lens of mateterialism's evolutionary theory. Hypothesis and narrative are always couched inside and never outside 'rationalism's' bio-case, a pitch based on the irrationality of aimless chance. *If, though, you've got chance, I've got logic; you've mindlessness, I mind. And isn't explanation based on reason actually the rational one?* Codes and programs, basic to biology, stand for reason, purpose and, rather than irrational evolutionary guesswork based on randomness, teleology.

---

[165] see Glossary: archetype; also *SAS passim.*

[166] *SAS* Chapters 22 and 23.

Holism can permit, of course, an immaterial element of information sourced in what we understand as mind; also specific coding, its potential for adaptation and the powerful but limited plasticity of archetypal variation. *But there's no need to buy the macro-evolutionary line.*

This book's critique has, as regards the fossils, fluid dates and ever-evolutionary interpretation, clearly shown muddle. So, has it cleared the smoke or readjusted mirrors helping you to see your origin more sharply? Did apes produce, for example great engineers, writers, prophets, scientists or philosophers? In this behavioural respect *top-down* decrees that to belittle the informative dimension is a superficial view; comparison between our '*Phylum cognoscens*' and chimps is intellectually cheap. Yes, there are similarities but man and ape are deliberately differentiated due to variations in their archetypal program; they embody preordained distinctions in routines and subroutines of regulation by their master and sub-magisterial genes. At this point, therefore, simply try to view the evidence from a *top-down*, dialectical angle. From here the 'anthropoid' condition is clear. *Of its set (new and old world monkeys, lesser apes and great apes) only great apes are of ape-man interest. This subset (of gorilla, orang-utan, chimpanzee and man) splits neatly into two - apes and men. No blur.* No fake parade, no finely-graded 'transitional' missing links, no communistic ape-man. *Man is man and ape is ape.* All that is found is one or the other. The distinction is archetypal; archetype is informative and information's source is metaphysical. The acceptance of metaphysical information as well as physical energy as a fundamental component of cosmos immediately invites a different perspective on the source of life and morality.

Final word. The origin of man's descent is not by way of randomness or even species. It is not by chance but by informative program.[167] Who informs any program or design but its informant; and, in the case of codified, human program, an Informant? *The problem is, bottom-up loudly, impatiently insists, that such a conclusion is not just theoretically improbable, it is **materialistically IMPOSSIBLE**....*

Of course it is!

---

[167] Descent of a program from idea to physical operation (*SAS* Chapter 6: The Order of an Act of Creation) is called vertical causation. Thus the vertical origin of man's descent is followed by the horizontal, knock-on operation of reproductive mechanisms and heredity (*SAS* Chapters 22, 24 and 25). These two forms of causation (explained in *SPFP*, Science and Philosophy - A Fresh Perspective) are not to be confused. They are as different in category as an inventor and his machine. In our case there exist two aspects of human origin and descent - the vertical and deliberate, original program along with its consequent horizontal, automatic and physical operation.

Fossil Catalogue Abbreviations

AL	Afar, Ethiopia
D	Dmanisi, Georgia
ER	East (Lake) Rudolf/ Turkana
KNM	Kenya National Museum
KP	Kanapoi, Kenya
KS	Kow Swamp, Australia
LH	Laetoli hominid
NG	Ngandong (Java)
OH	Olduvai hominid
SK	Swartkrans (South Africa)
Sts, Stw	Sterkfontein (South Africa)
WT	West (Lake) Rudolf/ Turkana

So, for example, fossil KNM-ER 1590 means a fossil catalogued by the Kenya National Museum found at a site in the East Rudolf area.

Hominoid Classification Reference

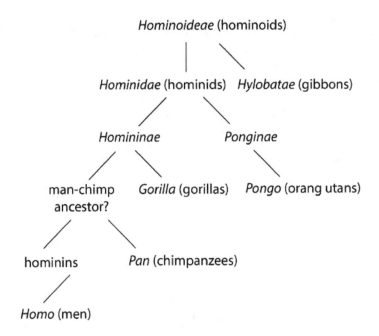

Uniformitarian (Evolutionary) Chronology

Era	Period	Epoch	Estimated Start millions of years (mya) ago
CENOZOIC	Quaternary	Holocene	0.01
		Upper Pleistocene	0.12
		Middle Pleistocene	0.8
		Lower Pleistocene	2.6
	Tertiary	Pliocene	5.3
		Miocene	23

Glossary (see also archaeologyinfo.com)

A

adaptive potential: involves pre-programmed, super-coded switches and recombinant (transposable) refinements intrinsic in the genomic program of any particular biological type (*SAS* Chapter 23).

allele: a sister gene; you have two copies of life's book, one from mother and the other from father, so that each gene from father has a correlate 'allele' from mother and *vice versa*.

allosome: sex chromosome (in humans X and Y); also heterosome.

anthropogeny: study of human beginnings in terms of evolution, culture etc.

anti-parallel perspectives: two anti-parallel perspectives substantiate the dialogue of Natural Dialectic; *top-down* implies you've got the information that you need - from 'on high', it is a system maker's expert point of view; *bottom-up* is taken as the empirical method of a humble student who, from child-like ignorance, starts from a clean state; such lack of preconception marks a strength of scientific method whose student learns by experiment; also implied in anti-parallel perspectives are *holistic* and *materialistic* faiths; the former involves both immaterial information (whose basis is always mind) *and* natural forces; the latter involves only material energy and thus invokes mindless, aimless behaviour as its sole creator of effects.

archetype: basic plan, informative element; conceptual template; pattern in principle; instrument of fundamental 'note' or primordial shape; causative information in nature; 'law of form'; nature's script; morphological attractor or subconscious field of influence in universal mind (see especially *SAS* Chapter 16); the subconscious component of universal (natural) mind comprising archetypes; prototype-in-mind (maybe related to Platonic ideas, Aristotelian entelechies and/ or Jungian archetypes) whose potential matter is seen as hard a metaphysical reality as, say, particles are physical realities; program(s) naturally stored in cosmic memory - simple in terms of inanimate physical 'law' (of particles and forces), complex in terms of animate structure/ function/ behaviour; information stored in a typical mnemone; in biology, metaphysical correlate of biological type/ super-species that is physically expressed in codified form as *adaptive potential*; metaphysical precursor; the collective unconscious of a type e.g. human type; as thought is father to the deed or plan is prior to ordered action, so archetypes precede physical phenomena; pre-physical initial condition of matter.

artefact:	something observed in a scientific investigation or experiment that is not naturally present but occurs as a result of the preparative or investigative procedure, that is, of experimental design; *archaeologically*, a cultural or technological object created by humans; archaeologists routinely identify human constructions without having to know their maker, motivation or function; they (and biologists) similarly identify bio-constructions (e.g. bones and fossils); however, neither artefact nor bio-construction (which involves complex, specifically-informative codes that give rise to a correspondent shape) are the product of natural forces and chance but of conceptual mind; we can recognise a bio-construction without having to know its maker, motivation or function; however, since it philosophically chooses to deny the existence of any natural, immaterial information, materialism's very life depends on attributing to bio-facts an 'appearance of design'; it is, the inference is forced to run, 'as if' mindless force and chance designed without design - a counterintuitive, non-commonsensical oxymoron conjured from a single word, bio-evolution.
autosome:	a non-sex chromosome (in primates any chromosome not X or Y).

B

base:	significant component of a *DNA* nucleotide: a letter in 'the book of life': there are 4 bases in the genetic alphabet - A (adenine), G (guanine), C (cytosine) and T (thymine): in the case of *RNA* base T is replaced by U (uracil).
base pair:	the conservative accuracy of genetic inheritance and the elegant construction of *DNA* are both dependent on a base-pairing rule viz. G pairs only with C and A with T (or U).
black box:	process or system whose workings are unknown.

C

centaur man	in Greek mythology a creature half man, half horse symbolising the polar state of flux between whose ($\uparrow$) higher, noble and ($\downarrow$) lower, bestial tendencies all humans are obliged to struggle through their moral lives.
chloroplast:	organelle in plant cells containing photosynthetic apparatus; contains its own, non-nuclear, circular *DNA* (ct-*DNA*) about 150,000 base pairs long.
chromatin:	a nuclear complex that, using histone proteins, helps package, reinforce and control the expression of genetic *DNA*.
chromosome:	a 'book' in the 'encyclopaedia' of life; the human genome contains 46 chromosomes.

cladistics: method of classification using diagrams called cladograms; organisms are collected into groups on the basis of shared (homologous) features; homologies are tallied and numerical rather than speculative, evolutionary/ phylogenetic links drawn up between organisms; cladism is thus a powerful, neutral, objectively detached tool of analysis and for this reason the technique enjoys growing popularity among the world's taxonomists.

code: the systematic arrangement of symbols to communicate a meaning; code always involves agreed elements of morphology (the form its symbols take), syntax (rules of arrangement) and semantics (meaning/ significance); without exception such prior agreement between sender (creator/ transmitter) and recipient involves intelligence.

codon: a 'word' in the genetic language; stands for an amino acid or a full stop; since more than one codon may stand for a single amino acid the genetic code is sometimes ill-perceived as 'degenerate'.

combinatorial space: a 4-digit lock has 10,000 combinatorial possibilities; an ordinary-sized 150-residue protein has 10^{159} such 'combinatorial spaces'; it has been estimated that the probability of a 'mutational trial' generating such a specific protein is 1 in 10^{77}. This, even allowing for several working possibilities, is the measure by which, in vast 'combinatorial space', 'gibberish' outweighs 'operational sense'; there exist perhaps 10^{80} particles in the entire known universe; you might, therefore, most reasonably assert (since you'd need other changes simultaneously to build a novel, complex protein, organelle or simple metabolic pathway) that sheer Mount Improbable cannot be mastered - not, at least, in gradual, evolutionary ways; Darwinian theory is locked right out of possibility; it is statistically dead.

conscio-material dipole: illustrates the basic components of polar existence; informative and energetic components may be graphically modelled (as *fig.* 2.5) on vertical y- and horizontal x-axes respectively; the origin of such graph (zero information and zero energy) represents the cosmic sink - an abyss of space); sources of the couple extend from infinity towards this sink; Archetypal Potential scales from Psycho-Logical, Informative First Cause through grades of mind to non-conscious zero (matter); and potential matter (first cause physical) drops from original concentration of ultra-heat to, again, zero; thus cosmos is viewed as the gradual embodiment of Uncreated Source; it represents a scale of possibilities expressed as typical yet individual forms; by this token embodied soul is subject to individual incorporations (psychological and physical); these relatively dynamic forms constitute its psychological and, in the biological case, bodily circumstance; Natural Dialectic simply models such a hierarchical description of polar creation by the use of spectrum, concentric rings and, step-wise, ziggurats (see also cosmos and creation below).

cosmos: often applied to physical universe, universal body; from Greek word meaning orderly as opposed to chaotic process; involuntary pattern of nature; also equated, including metaphysical mind, with existence as a whole; seen, dialectically, as a projection through the template of metaphysical archetypes; **the umbrella title of the series and website of books, Cosmic Connections, could, with reference to the Natural Dialectic which structures its *CUT* (see Glossary: unification), equally be called Orderly Linkages.**creation:origination; physical or psychological arrangement; mind creates with purpose, matter without; creation means active production but also passive result; a creation will have been informed by force of mind and/or matter**creation:** origination; physical or psychological arrangement; mind creates with purpose, matter without; creation means active production but also passive result; a creation will have been informed by force of mind and/or matter.

creationist: one who believes in creation of the universe and living bodies; a term perjoratively used by materialism as a philosophical weapon to sustain its world-view; such ridicule specifically targets a literal belief in the biblical Book of Genesis; in general terms materialists/ anti-creationists (*de facto* members of the evolutionary faith) believe there is no immaterial element i.e. no natural element of information, reason or symbolic capacity; anti-creationism believes the atoms of brain together produce, in varying degrees, intelligence in the form of knowledge and memory; if materialism self-confuses with regard to creative potential it certainly refuses Creativity or a Creator; such anti-creationists thereby categorically reject purpose in nature, holism, all religious faith, the point of contemplation, the heart of meditation and, logically, even the essential, immaterial, conscious character of their own minds (see *SAS passim*). Creationists and anti-creationists do not see anything, except material effects, the same way.

D

diploid: having full genetic complement with one copy of chromosomes from each parent e.g. you have 46 chromosomes, 23 from mother and 23 from father.

DNA: a complex chemical; a large bio-molecule made of smaller units, nucleotides, strung together in a row; a polymer in the form of a double-stranded helix; a medium superbly suited to the storage and replication of 'the book of life'; paper and ink' on which the genetic code is inscribed; an organism's 'hard drive'.

E

entropy: a measure of the amount of energy unavailable for work or degree of configurative disorder in a physical system (see second law of thermodynamics); inertial aspect of an energetic, material or conscious gradient; diffusion or concentration gradient outward from source to sink; drop towards 'most probable' outcome i.e. inertial slack; a measure of disintegration or randomness; expression of the (*tam* ↓) downward cosmic fundamental; a major property of matter, closely coupled with materialisation; in a closed system, which the universe may or may not be, this tends the eventual loss of all available energy, maximum disorder and the exhaustion of so-called 'heat death'.

enzyme: protein catalyst without whose type metabolism (and therefore biological life) could not happen.

epigeny: genetic super-coding; contextual punctuation; chemical modification of *DNA*; also extra-nuclear factors that may cross-reference with genetic expression; total epigeny is vested in the epigenome.

Essence: (*Sat*) Supreme or Infinite Being; Substance (perhaps Spinoza's Substance) 'prior to' or 'above' existence; Pure Consciousness/ Life; Peace that transcends all psychological and physical action; the root of an essentially undivided universe; Uncreated One within which and whence all differences have their being; Apex of Mount Universe; goal of saints/ 'philosopher kings'; the 'point' at which All-Is-One.

eukaryote: non-prokaryote; any organism except bacteria and blue-green algae.

evolution: there are today *four* main usages of this word; each 'loading' derives from the original Latin, 'evolvere', meaning to unroll, disentangle or disclose; the *first two*, physical and biological, are conceived as natural/ mindless processes; the *second, mindful pair* is of psychological/ teleological import; specious ambiguity may conflate or switch between the fundamentally separate pairs of meaning. **Firstly,** in the scientific context of physics and chemistry, the word is used to describe change occurring to physical systems; the laws of nature can't, it seems, evolve through time but stars, rocks and gases can. **Secondly**, though also subject to the 'rules' of entropy, biological evolution is a theory of *random progression* from simple to complex form; it thereby implies increasing, codified complexity; while retaining the 'hard loading' of physical science it also, ambiguously, claims that codes, programs, mechanisms and coherent, purposive systems - normally the province of mental concept - self-organise by, essentially, chance; such confusion, the basis of naturalism, is compounded by failure to distinguish between, on the one hand, ubiquitously observed variation (called micro-evolution) and, on the other, Darwinian 'transformation' between different sets of body plan, physiological routines and associated types of organism - such 'black-box macro-evolution' as is never indisputably observed; to evoke a naturalistic ambience it is fashionable to use 'evolved' interchangeably with or to replace the words 'was created', 'was planned' or 'designed'; finally, it is noted that the coded, choreographed development of a zygote, packed with anticipatory information, through precise algorithms to adult form is the absolute antithesis of blind Darwinian evolution. *Thirdly*, man certainly evolves ideas; intellect can evolve 'purposive complexity'; we invent all kinds of codes, schemes and machines; we devise increasingly complex theories and technologies; and we evolve an understanding of natural principles; this, which all parties accept, is an informative, psychological sense of 'evolution'. The *fourth* sense of evolution, at least as near to the original Latin as the other three, is the spiritual usage; immaterial spiritual evolution, unacceptable to materialists and unknown to physical science, is at the very heart of holism; in this voluntary sense of evolution practitioners cast off material attachment, evolve and merge into the *Logos*; evolution (or, perhaps better, centripetal involution) of the soul is their great business; their aspiration is to unite with The Heart of Nature.

evolution pre-Darwinian: minority/ anti-mainstream pre-Socratic snippets and sense-based Epicureanism lionized by interpretations of post-18th century materialists; virtually undetectable eccentricity in Chinese, Indian and Islamic literature; natural selection treated by creationists al-Jahiz and Edward Blyth; Buffon, a non-evolutionist, addressed 'evolutionary problems'; Lamarck (evolution by inheritance of acquired characteristics); hints in poem by Erasmus Darwin.

evolution Darwinian: mechanism - natural selection; major tenets - common descent (inheritance), homology and 'tree of life'.

evolution: neo-Darwinian/ synthetic: as Darwinian, except synthetic theory adds random mutation as the mechanism for innovation; also adds a mathematical treatment of population genetics and various elements (e.g. geno-centric perspective) derived from molecular biology.

evolution: post-synthetic phase: natural selection and random mutation are acknowledged as mechanisms insufficient to source bio-information; post-Darwinian evolution invokes mechanisms from hypotheses such as *NGE* (natural genetic engineering) and 'evo-devo'; holistic possibilities also address the origin of complex, specified and functional bio-information.

existence: which 'stands out' from background 'nothingness'; the apparently divided universe; seemingly disparate, finite things; all motion/ change/ relativity; all psychological and physical events.

exon: specifies the amino acid sequence for a protein; m-*RNA* after protein editors have removed introns.

G

gamete: sex cell with half of full genetic complement i.e. a single set of chromosomes.

gene: generally means a basic unit of material inheritance; section of chromosome coding for a protein; digital file; a reading frame that includes exons and introns; the old one gene-one protein hypothesis is incorrect; in fact, by gene splicing, a particular piece of *DNA* may be used to create multiple proteins.

genetic drift: fortuitous change in frequency of an allele in a population.

genome: total genetic information found in a cell: think of the genome as an instruction manual for the construction and physical operation of a given organism.

genotype: the genetic constitution of an organism, often referring to a specific pair of alleles; the prior information, potential, plan or cause of an effect called phenotype.

H

haplogroup: a group of haplotypes that share SNPs (see SNP) and, thereby, an assumption of common ancestry; haplogroups used to define descent are ones that are unaffected by sexual recombination, that is, patrilineal Y-chromosomal amd matrilinear mitochondrial. Changes are, therefore, by chance alone. There are a number of haplogroups but the oldest matrilinear group (L0) is found in Khoisan tribes of South Africa. The oldest of all (A00) belongs to a group of people co-existing in Africa, Europe, and Asia; it includes multiple human types, including *H. sapiens, H. erectus, H. heidelbergensis*, and even Neanderthals.

haplotype: a set of *DNA* variations inherited unchanged or involving only a specific 'single nucleotide mutation' (SNP).

haploid: having half the full genetic complement, as in the case of sex cell.

heterochrony: change in developmental timing or rate leading to changes in size or shape (e.g. antler, skull or tail size).

heterosome: sex chromosome (in humans X and Y); also allosome.

heterozygous:	having different allelic forms of a particular gene.
holism:	opposite of reductionism; the view that a whole is greater than the sum of its parts; the extra metaphysical (immaterial) ingredient is identified by Natural Dialectic as information; information implies the purposeful design, development and arrangement of contingent parts in a working system; may operate according to a Logical Norm.
homeostasis:	vibratory or periodic control of a system to obtain balance round a pre-set norm; the mechanism of its information loop involves sensor, processor and executor; the operative cycle works by negative feedback; psychological (nervous) and biological cybernetics; the informed basis of biological stability.
homeotic gene:	gene involved in developmental sequence and pattern; high-level co-determinant of the formation of body parts.
hominid:	member of family *Hominidae*; first hominid unknown; confusingly, the term 'hominid' used to mean what 'hominin' does now; thus 'hominid' can mean two different things according to the text used.
Hominidae:	great apes (human and extinct relatives, chimpanzee, gorilla and orang-utan): confusingly, until 1980 *Hominidae* meant human and extinct relatives only while non-human great apes were called *Pongids*).
hominin:	modern or extinct humans: first hominin unknown.
Homininae:	great apes (human and extinct relatives, chimpanzee and gorilla).
Hominini:	great apes (human and extinct relatives).
hominoid:	member of *Hominoideae*.
Hominoidea:	superfamily composed of great apes (human and extinct relatives, chimpanzee, gorilla, orang-utan and gibbon).
homozygous:	having the same allelic forms of a particular gene.
humanoid:	non-scientific term meaning of human appearance and/ or character (thereby including at least *hominin*).

I

illusion:	is the cut between illusion and delusion an illusion? illusions, apparently outside the mind, appear real; a delusion, in it, we think real; neither, mind allows, is real or true.
indel:	insertion and/ or deletion of *DNA* bases (as opposed to substitution or replacement called a point mutation).
information:	the immaterial, subjective element; information is the inhabitant of its own centre, mind, whose substrate is consciousness; active information knows, feels, purposes and codifies; it recognises meaning; on the other hand passive information reflects active; it is stored as subconscious memory; or is fixed in the expressions of non-conscious matter according, universally, to the

archetypal behaviours of natural bodies or, locally, to particular constructions by life-forms.

information entropy: loss of information due to degradation of its carrying medium; such a medium may be active and metaphysical (mind) or passive and physical (for example, computer files or genetic code); and such entropy may be metaphysical (loss of memory, focus or consciousness) or physical (for example, genetic mutation); the informative correlate of such degeneration is diminished organisational capacity, meaning or thrust of original purpose.

intron: genetic control panel; n-p-c (non-protein-coding) segment(s) spliced from an m-*RNA* transcript prior to translation; introns include regulatory elements (to variably promote or inhibit gene expression) and addressing factors of the genetic operating system; gene-attached information lending specific flexibility to protein manufacture.

L

Lamarckism: evolution by inheritance of acquired characteristics.

logic: analysis of a chain of reasoning; principles used in circuitry design and computer programming; 'normative reason' relates to the basic axiom(s) of a given standard e.g. *bottom-up* materialism or *top-down* holism; three main logical thrusts are: (1) inductive (premises/ observations supply evidence for a probable/ plausible conclusion) as in the case of experimental science working *bottom-up* from specific instances to general principle: (2) abductive (best inference concerning an historical event): and (3) deductive (conclusion in specific cases reached *top-down* from general principle): two pillars of logic are holism and materialism; holism employs mainly deductive/ abductive operations and a Logical Norm; materialism tends to inductive/ abductive operations whose axis is non-conscious force and chance.

Logos: First Cause; Prime Mover; Causal Motion that sustains creation's conscio-material gradient.

M

macro-evolution: large-scale, non-trivial evolution; process of common or phylogenetic descent alleged to occur between biological orders, classes, phyla and domains; includes the origin of body plans, coordinated systems, organs, tissues and cell types; unexplained by mutation, saltation, orthogenesis or any known biological mechanism; sometimes called 'general theory of evolution' (*GTE*); macro-evolution, an extrapolation from Darwinian micro-evolution vital to sustain the materialistic mind-set, is conjecture.

meiosis: shuffling the information pack: variation-on-theme; mechanism for the production of haploid gametes; genetic postal system for sexual reproduction.

208

metabolism: body chemistry.

metaphysic: = non-physical/ immaterial/ psychological/ unnaturalistic; physically expressed as specific/ intended arrangement/ behaviour of materials; physical behaviour reflects metaphysical blueprint; involves element of information; involves symbol/ code/ abstraction/ logic/ reason/ mathematics; involves meaning/ message/ goal/ teleology; involves consciousness/ mind/ life/ experience/ feeling; involves morality/ force psychological/ emotion; involves innovation/ creativity/ art/ invention/ aesthetics.

micro-evolution: misnomer; non-progressive, small-scale variation within Linnaean classifications of strain, race, species or, more broadly, genus and perhaps even, on occasion, family; limited plasticity; variation/ adaptation within *type*; trivial Darwinian changes that may occur by natural selection/ ecological factors acting on genetic recombination or mutation; sometimes called 'special theory of evolution' (*STE*), micro-evolution/ variation is a fact.

mitochondrion: organelle in eukaryotic cells containing the apparatus for aerobic respiration; contains its own, non-nuclear, circular *DNA* (mt-*DNA*) about 16000 base pairs long.

mitosis: conservative copying and delivery of genomes in cell division; genetic reprinting; genetic postal system for asexual reproduction.

mobile genetic element: transposon, retrotransposon, insertion sequence, other non-protein-coding *DNA*, n-p-c *RNA* fragments and various protein regulators that together expedite the operating system of a genome.

morphogenesis: the development of biological structure; more generally, the production of physical form.

mosaic: the presence of permutations of sub-routines or similarities of form and/or function scattered in organisms unrelated by lineage.

mutation: accidental change to genetic code.

N

nano-biology: biology of structures/ physiologies involving a few atoms or molecules; 'extremely small biology'.

nano-technology: technology at atomic and sub-atomic level as is, basically, life's.

naturalistic methodology: also known as 'methodological naturalism': is, strictly, not concerned with claims of what exists or might exist, simply with experimental methods of discovering physically measurable behaviours; thus only material answers to any question (e.g. how biological forms arose) are deemed 'scientific' or 'scientifically respectable'.

negentropy: opposite of entropy; lowering of entropy; expression of the (*raj*) upward-pointing cosmic fundamental closely coupled with stimulus, dissolution and dematerialisation; a measure of input, cooperation or synthesis; motive/ fluidising aspect of an energetic, material or conscious gradient; gain of energy, configurative order, information or consciousness in a system; when used in terms of information negentropy involves gain in order or understanding of principle from which different actualities derive; a measure of the amount of concentrated/ conceptual information, specific, intentional complexity or conscious arrangement in a system; a natural and essential property of mind.

non-protein-coding *DNA*: occupies probably 95% of eukaryote and 80% of bacterial genomes; associated with the genetic operating system; may include some genuinely redundant misprints or duplications but now thought for the most part critical to the flexibility, efficiency and even possibility of gene expression; once thought of as useless, degraded information and ignorantly called 'junk *DNA*'.

non-protein-coding *RNA*: n-p-c *RNA* is also called n-c-*RNA* (non-coding), nm-*RNA* (non-messenger) or f-*RNA* (functional); functional *RNA* molecule not translated into protein; many 1000's of different specimens include classes of t-*RNA* (transfer *RNA*), r-*RNA* (ribosomal *RNA*) and, commonly involved in the regulation of gene expression and other intra-cellular tasks, micro-*RNA*, double-stranded si-*RNA*, pi-*RNA* and so on; also, for inter-cellular communication, ex-*RNA*.

nucleic acid: *see DNA and RNA*

nucleotide: basic, triplex unit of nucleic acid polymer; monomer composed of phosphate and sugar (the 'paper' part) and base (the 'ink letter'); letters' of the genetic alphabet are (G) guanine, (C) cytosine, (A) adenine and (T) thymine. In *RNA* thymine is replaced by (U) uracil.

O

order: regular, regulated or systematic arrangement; organisation according to the direction of physical law; passive information by which things are arranged naturally (with predictable but non-purposive complexity) or purposely (with innovative or specified complexity); mind, generating specified complexity in the order of its technologies and codes, actively informs; the orders of mind are meaningful, the orders of matter lack intent; see also cosmos.

organelle: cellular sub-station; discrete part of a cell; sub-cellular compartment having specific role such as informative (nucleus), energetic (mitochondrion, chloroplast), constructional (ribosome, Golgi body) or other.

P

palaeoanthropology: the study (by gene or fossil) of human evolution.

palaeontology: the study of prehistoric life and fossils.

PAM, PAND, PCM **and** *PCND*: philosophical gambits; see Primary Axioms and Corollaries.

phenotype: the effect of causal potential; result of the development of prior, informative 'egg'; outward expression of inner plan; sensible appearance of an organism as opposed to its genotypic scheme: the whole set of outward appearances of a cell, tissue, organ and organism are sometimes called a phenome (*cf.* genotype/ genome).

phylogeny: evolutionary history; relationships based on common or evolutionary descent.

polygenism: proposition that human races come from different ancestors (no single 'Adamic' stock).

Primary Axiom of Materialism (*PAM*): all objects and events, including an origin of the universe and the nature of mind, are material alone; cosmos issued out of nothing; life's an inconsequent coincidence, a fluky flicker in a lifeless, dark eternity.

Primary Axiom of Natural Dialectic (*PAND*): there exists a natural, universal immaterial element - information; immaterial informs material behaviour; a conscio-material dipole that issues from First Cause informs and substantiates both mental (metaphysical) and physical creations; there is eternal brilliance whose shadow-show is called creation.

Primary Corollary of Materialism (*PCM*): the neo-Darwinian theory of evolution, that is, life forms are the product, by common descent, of a random generator (mutation) acted on by a filter called natural selection; such evolution is an absolutely mindless, purposeless process; the *PCM* is a fundamental *mantra* of materialism.

Primary Corollary of Natural Dialectic (*PCND*): the origin of irreducible, biological complexity is not an accumulation of 'lucky' accidents constrained by natural law and death; forms of life are conceptual; they are, like any creation of mind, the product of purpose.

prokaryote: non-eukaryote; bacterial type with little or no compartmentalisation of cell functionaries.

promissory materialism: belief sustained by faith that scientific discoveries will in the future justify/ vindicate exclusive materialism and, as a consequence, atheism; may involve a call to progress towards the technological provision of its 'promised land'.

protein: factor made from a specific sequence of amino acids to perform a specific task; 'informative' protein includes some hormones; skin, hair, bone, muscle and other tissues are made of 'structural' protein; 'functional protein' called enzymes mediates all stages in cell metabolism, that is, it catalyses all biochemistry.

___

R

reductionism: opposite of holism; the materialistic view that an article can always be analysed, split up or 'reduced' to more fundamental parts; these parts can then be added back to reconstruct the whole; a whole is no more than the sum of its parts.

respiration: the controlled release of energy from food.

ribosome: site of polypeptide (protein) synthesis.

RNA: a single-stranded nucleic acid polymer employed in three different forms during the process of protein synthesis; in computer terms might be likened to a portable memory stick as opposed to *DNA*'s hard drive.

m-RNA is used to transcribe a base sequence from *DNA*. It 'photocopies' a gene and carries this information to a ribosome.

micro-RNA short mi-*RNA* molecules are important regulators of genetic expression.

r-RNA is part of the make-up of the protein-manufacturing station called a ribosome.

t-RNA critically translates genetic 'words' (see 'codon') into amino acids: 64 such operators form the link between code and the actuality of a functional protein.

___

S

SAS: Science and the Soul (see Bibliography).

satellite DNA: repetitive non-protein-coding *DNA* (also called a variable number tandem repeat); main component of centromeres and telomeres; constituent of tightly packed chromatin affecting the regulation of gene expression.

science: Latin *scire* (know); knowledge; commonly understood as the practical and mathematical study of material phenomena whose purpose is to produce useful models of the physical world's reality.

scientism: a philosophical face of official, *de facto* commitment to materialism; today's majority consensus of what the creed of science is; an -ism born of *PAM*; a faith that all processes must be ultimately explicable in terms of physical processes alone; like communism, a one-party state of mind; a doctrine that physical science with its scientific method is ultimately the sole authority and arbiter of truth; a set of concepts designed to produce exclusively material explanations for every aspect of existence, that is, to colonise each academic discipline and build its intellectual empire everywhere; 'scientific fundamentalism' closely allied, when expressed in social and political terms, with 'secular fundamentalism', sociological interpretation of behaviour and the fostering of a humanistic curriculum.

secular fundamentalism: *PAM* as applied to the worlds of nature and of human society.

secularism: concern with worldly business; lack of involvement in religion or faith; secularism is generally identified, as defined by the dictionary, with materialism; for a secularist the ultimate arbiter of truth is human reason - ideas are open to negotiation so that even morality is relative; however many liberal agnostics, atheists and humanists argue that their metaphysical, philosophical system also embraces so-called 'universal' moral values and, as opposed to zealotry or the logic of evolutionary faith, a liberal politic of 'philosophical live-and-let-live'.

SNP: a particular sort of mutation; a *functional SNP* or 'single nucleotide polymorphism' consists of a single base change in a *DNA* sequence that affects a factor such as gene splicing or messenger *RNA*; a *coding SNP* occurs within a gene and, whatever the interpretation placed upon its observed effect may be, is conserved and therefore probably neutral or useful.

T

taxon: a group of one or more populations of an organism or organisms judged to form a taxonomic unit.

taxonomy: biological science of description, identification, nomenclature and classification of organisms; species of taxonomy include Linnaean binomial, phylogenetic (evolutionary) and cladistics (grouping by shared characteristics).

teleology: the doctrine that there is evidence of purpose in nature; doctrine of non-randomness in natural architecture; doctrine of reason ('for the sake of', 'in order to', 'so that' etc.) and intent behind biological and universal design.

transcendent projection: orderly, energetic expression from either metaphysical or physical nothingness, that is, unseen potential; instantaneous miracle from 'beyond' non-conscious physicality; transcendently emergent, finely tuned expansion from 'inner' metaphysic into 'outer' material/ natural law; 0-dimensional singularity from whose prior pointlessness all points perhaps began; cosmic seed whence, *ex nihilo*, the world developed; projection whose physical appearance is perhaps described but not explained by big bang theory; perhaps an ultimately incomprehensible dynamic.

transposon: 'jumping gene'; ubiquitous genetic element found in all prokaryotes and eukaryotes so far investigated; *DNA* segment that can, by enzyme, be cut from a one site (the donor) and joined to another (the target); a retrotransposon is moved through the mediation of *RNA* and reverse transcription back from *RNA* to *DNA*; a kind of retrotransposon, *SINE*s and *LINE*s are thought to compose 30+% of the human genome; from an evolutionary view they comprise functionless viral imports; from a *top-down* view it is predicted they will be found to form a dynamic, intrinsic element of the genome involved in gene regulation, genetic shuffling as (epigenetic) response to buffer circumstantial exigency and, as important, structural agents able to reshape a chromosome to meet specific genetic demands.

U

unification: simplification; details are unified by their working principles, themes or programs; better to perceive intrinsic principle is to simplify or unify an understanding; progressive unification of forces is the grail of physics: Clerk Maxwell unified electricity and magnetism; electroweak or *GSW* theory brought in the weak nuclear force; now the goal is to include the strong nuclear force (*GUT*), gravity in a super-force and show that, in essence, particles and forces are interchangeable (super-symmetry and *TOE*); Natural Dialectic, also working with the maxim 'All is One', includes what sums to a hierarchical *TOP* or Theory of Potential; potential (see *SAS*: Glossary and Index for archetype and potential) is an absolute from which variant orders of relativity derive; the equivalent of *TOP* is *CUT* (see index: cosmic unification theory); **Natural Dialectic is the vehicle of *CUT*, whose aim is to build cosmic connections, that is, orderly linkages towards a Holy Grail of Unification**; the Great Connector, that is, Unifier is consciousness; the subjective potential for mind is consciousness and the objective potential for matter is archetypal memory; such archetypal element unites psychology with the physics of natural science; it is the informative precondition of physical and biological form.

universal mind: cosmic grade; also called the 'mind of nature' or 'natural mind'; as a biological body is a specific though complex arrangement of universal chemicals so individual mind partakes of a particular, equally minuscule fraction of the metaphysical components of universal mind; *see* also archetype.

X

X-chromosome: XX is associated with femaleness (♀) in mammals.

Y

Y-chromosome: associated with male sex (♂) in mammals.

Z

zygote: fertilised egg.

Index

Bibliography

Adam and Evolution	Pitman M.	2015
Ape Men Fact or Fallacy?	Bowden M.	1977
Bones of Contention	Lubenow M.	2008
Charles Darwin: Victorian Mythmaker	Wilsin A.N.	2017
Darwin on Trial	Johnson P.	1991
Darwin's Black Box	Behe M	1996
Darwinian Fairytales	Stove D.	1995
Descent of Man	Darwin C.	1871
Devil's Delusion	Berlinski D.	2009
Evolution, The Human Story	Roberts A.	2011
Evolution's Achilles Heels	ed. Carter R.	2014
God: To Be or Not To Be	Wilder Smith A. E.	1975
Icons of Evolution	Wells J.	2000
Macro-evolution	Stanley S.	1979
Natural Sciences Know Nothing of Evolution	Wilder Smith A.E.	1981
Natural Theology	Paley W.	1802
One Small Speck to Man	Sodera V.	2009
Origin of Species	Darwin C.	1859
Reason in the Balance	Johnson P.	1995
Rocks Aren't Clocks	Reed J.	2013
Science and Human Origins	Gauger, Axe, Luskin	2012
Science and the Soul	Pitman M.	2015
The Origin of Our Species	Stringer C.	2012
Y, The Descent of Men	Jones S.	2002

Image Credits

5.2 1. en.wikipedia.org CC0 1.0 Uniiversal Public Domain
  2. photo by and courtesy of Roberto Saez (2014)
  3. en.wikipedia.org CC BY-SA 3.0: author Nrkpan
  4. as drawn for Kenneth Oakley
  5. as drawn for British Museum (Natural History)
  6. as drawn for National Geographic 9-60
  7. commons Wikipedia.org CC BY-SA 3.0: author Lilliyundfreya
  8. as drawn for The Times 5-4-64
5.3 low resolution copy of Zaillinger illustration; Terry Picton
5.4 fair use: evolution of evolutionary stories (also cover)
5.5 courtesy of Discovery Institute
8.1 1. public domain: author Ernst Haeckel Anthropogenie (1874)
  2. en.wikipedia.org/wiki/image:user Eileen McGinnis
10.1 en.wikipedia.org: public domain
14.2 Neandertaler Schädelkalotte Inv.Nr. 322_1: Foto: Jürgen Vogel
  LVR - LandesMuseum Bonn
14.3 1. en.wikipedia.org CC BY-SA 3.0: author AquilaGib
  2. en.wikipedia.org CC BY-SA 2.5: author Utilisateur-120
  3. en.wikipedia.org CC BY-SA 2.0: author Jose Luis Martinez Alvarez
15.1 1. en.wikipedia.org PD-old, PD-US: author Personal Scan 120
  2. public domain: author Gabriel von Max (1894)
15.2 en.wikipedia.org CC BY-SA 3.0: author unknown
16.1 1. © The Trustees of the Natural History Museum, London
  2. en.wikipedia.org public domain: author John Cooke (1915)
  3. Wellcome Images M0013579: CC-Y 4.0.
16.2 en.wikipedia.org public domain: author Popular Science Monthly (1913)
17.1 commons Wikipedia.org CC BY-SA 3.0: author zoals
17.2 1. Andy English: wood engraver, illustrator
  2. Andy English: wood engraver, illustrator
  3. Wellcome Images M0001113: CC-Y 4.0
18.2 courtesy of DK Press
18.3 Yale Peabody Museum of Natural History
18.4 courtesy of The Georgian Museum
19.2 Wellcome Images L0067078: CC-Y 4.0
19.3 1. Wellcome Images L0067081: CC-Y 4.0
  2. en.wikipedia.org CC BY-SA 2.5: author Jose-Manuel Benito Alvarez
19.4 en.wikipedia.org CC BY-SA 3.0: author Nadina
19.5 en.wikipedia.org CC BY-SA 3.0: author Dr. Gunter Bechly
19.6 courtesy of Bradshaw Foundation
19.7 en.wikipedia.org CC BY-SA 2.5: author Luna04~commonswiki assumed
19.8 en.wikipedia.org GFDL (GNU free doc. licence) San Diego Mus. of Man soerfm
19.9. en.wikipedia.org public domain: author Jose-Manuel Benito Alvarez
19.10 wikimedia commons Nicolas Perrault III CC0 1.0; art prehistoric wiki CC BY-SA
20.2 1. Wikimedia commons: author Locutus Borg
  2. en.wikipedia.org CC BY-SA 3.0: author Didier Descouens

    3. North Carolina School of Science and Mathematics, licensed CC BY-NC-SA 4.0. Used with permission.

20.3 en.wikipedia.org CC BY-SA 4.0: author Didier Descouens

20.4 1. en.wikipedia.org CC BY-SA 3.0: author Ghedoghedo
    2. courtesy of Andrea Sforzi, Maremma Natural History Museum

20.5 1. en.wikipedia.org CC BY-SA 3.0: author it:Use:Lucius
    2. fair use: semi-humanised creature drawn from fragments of (mostly leg) bones

20.6 en.wikipedia.org CC BY 2.0: author T. Michael Keesey

20.7 en.wikipedia.org CC BY-SA 2.5: author 120

20.8 1. wikipedia wikia familypedia CC-BY-SA 3.0 fair use
    2. en.wikipedia.org CC BY-SA 3.0: author Jlorenz1

20.9 en.wikipedia.org CC BY-SA 4.0 international: photo by Brett Eloff courtesy Prof. Berger and Wits University who release it under GNU free documentation licence.

20.10 en.wikipedia.org CC BY-SA 3.0: author Bjørn Christian Tørrissen

21.3 en.wikipedia.org CC BY-SA 3.0: author Nachosan

21.4 en.wikipedia.org CC BY-SA 3.0: author Conti

21.5 fair use: image courtesy of Manila Bulletin

21.6 en.wikipedia.org CC BY-SA 2.0: author Ryan Somma

21.7 1. en.wikipedia.org CC BY 4.0: author Lee Roger Berger research team
    2. commons Wikipedia.org CC BY-SA 4.0: author Berger et al. 2015
    3. courtesy Antoine de Ras

23.1 fair use: image from 'Adam and Eve (The True Story)' issued under Standard YouTube Licence

23.2 courtesy of Bradshaw Foundation

24.1 1. en.wikipedia.org. public Domain: designed by Carl Sagan and *Frank* Artwork prepared by Linda Salzman Sagan. Photograph by NASA Ames Resarch Center Drake
    2. en.wikipedia.org. CC BY-SA 2.0: author Dr Mike Baxter

24.2 wikimedia commons CC-BY-SA 3.0; *H. longi* © N. Tamura

24.3 en.wikipedia.org public domain: author Hieronymus Wierix after Albrecht Durer

The author has recently written a few more books (available from Amazon, Foyles, Waterstones, Barnes & Noble etc. and see website addresses on p.2):

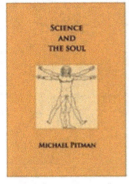

SCIENCE
AND
THE SOUL

MICHAEL PITMAN

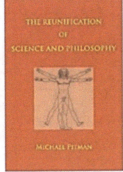

THE REUNIFICATION
OF
SCIENCE AND PHILOSOPHY

MICHAEL PITMAN

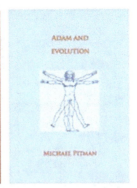

ADAM AND
EVOLUTION

MICHAEL PITMAN

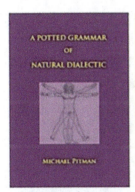

A POTTED GRAMMAR
OF
NATURAL DIALECTIC

MICHAEL PITMAN

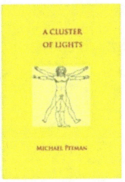

A CLUSTER
OF LIGHTS

MICHAEL PITMAN

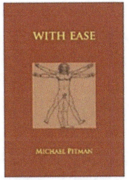

WITH EASE

MICHAEL PITMAN

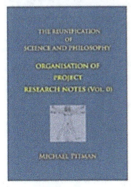

THE REUNIFICATION
OF
SCIENCE AND PHILOSOPHY

ORGANISATION OF
PROJECT
RESEARCH NOTES (VOL. 0)

MICHAEL PITMAN

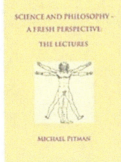

SCIENCE AND PHILOSOPHY -
A FRESH PERSPECTIVE:

THE LECTURES

MICHAEL PITMAN

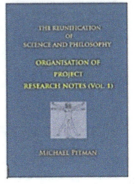

THE REUNIFICATION
OF
SCIENCE AND PHILOSOPHY

ORGANISATION OF
PROJECT
RESEARCH NOTES (VOL. 1)

MICHAEL PITMAN

www.ingramcontent.com/pod-product-compliance
Lightning Source LLC
LaVergne TN
LVHW012329060326
832902LV00011B/1792